HANDS-ON METEOROLOGY

Stories, Theories, and Simple Experiments

by

Zbigniew Sorbjan

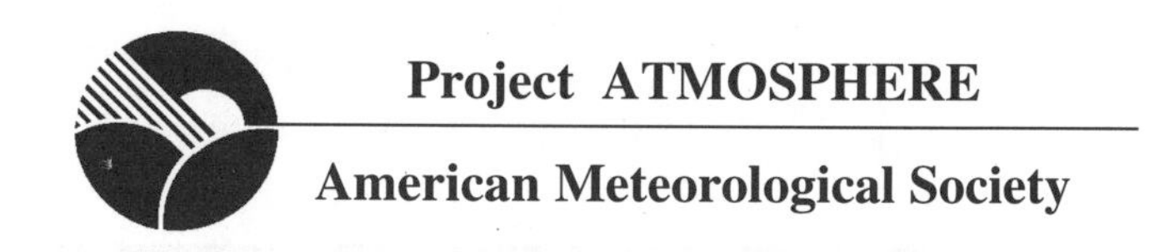

First Edition, 1996

Cover design, layout design, and illustrations by Zbigniew Sorbjan

Library of Congress Cataloging-in-Publication Data

Sorbjan, Zbigniew
HANDS-ON METEOROLOGY
Stories, Theories, and Simple Experiments

p. cm.
Includes bibliographical references and index.
ISBN: 1-878220-20-9
1. Meteorology — Popular works. I. Title.

To Grace, to Zosia,
and in the loving memory of
my mother, Zofia Sorbjan

CONTENTS

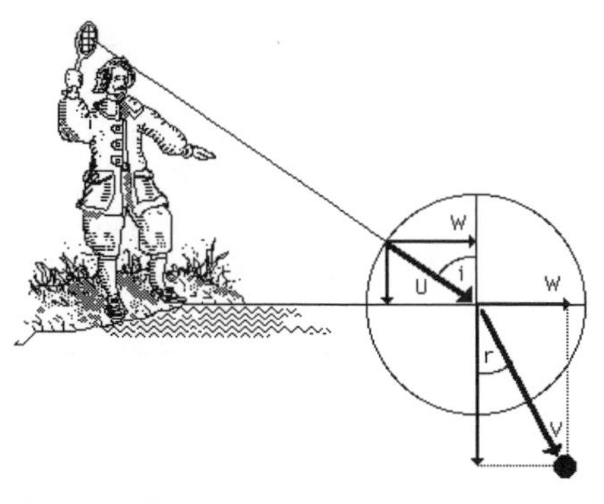
W
U
i
W
r
V

Rough is the road that leads to the heights of greatness. (Seneca)

PREFACE

In 285 B.C. the Egyptian King, Ptolemy, engaged the services of the well-known mathematician, Euclid, as his tutor. After considerable mental anguish, the King asked his teacher if he could provide an easier method of mastering the subject. Euclid uttered his now famous reply: "There is no royal road to geometry."

The moral of the above story also applies to meteorology: teaching it and learning about it are intellectually challenging. Throughout the years, countless numbers of teachers and students have echoed those same thoughts. Therefore, this book is a result of my intention, as both a researcher and a teacher, to present the subject of atmospheric science in an exciting, intriguing, and even entertaining way.

As indicated in the title, the book utilizes a "hands-on" approach by allowing readers to perform simple experiments. Since the focus of the book is the atmosphere, the selected experiments demonstrate concepts of basic meteorology: air pressure, heat, temperature, moisture, and wind. The experiments are easy to do and designed to discover understandable explanations for a variety of questions, and also to develop an interest in the habits of critical, scientific and creative thinking. Many of the materials used in the experiments are common, everyday, inexpensive ones such as candles, bottles, cans, and balloons.

Ever mindful of safety, especially for younger experimenters, the author alludes to safety. Although all of the experiments detailed in this book are safe, standard laboratory precautions are always recommended. For example, experimenters are cautioned to be careful when handling hot glass -- a material which does not look hot and cools very slowly! Participants are also told to be cautious with hot plates, burners, candles and matches. They should also avoid pointing heated test tubes at others and are warned about the dangers of tasting and smelling different chemicals.

"Hands-On Meteorology" does not deal exclusively with meteorology or meteorological experiments; it also contains numerous historical narratives. Furthermore, it includes references to important cultural events and famous, as well as infamous, individuals. Interesting sketches about the lives and works of prominent scientists, along with a number of meteorological anecdotes and illustrations, should prove entertaining and capture the interest, fancy, and imagination of readers of all ages.

Another feature of this book is that it varies in difficulty. The reader can decide on skipping sections which appear too trivial, or too technical. Yet, a continuity remains in the subject matter as a whole. Sections in each chapter are short and provide an independent completeness where sequential reading is necessary. Detailed information on the "hows, whys, wheres, and whens" enables readers to understand facts through discovery and enjoyment. The main topics in this book are arranged in chronological order, wherever possible. For those who wish to become more deeply involved in the subject, references are provided at the end of the book.

In essence, "Hands-on Meteorology" has been addressed to all those who are enthusiastic about learning and teaching about the atmosphere, weather, and climate. And my contention has been to show that after all there might be a "royal road to meteorology".

I would like to thank all of those who have assisted me in the preparation of this book. I am indebted to Dr. Edward J. Hopkins, Dr. Richard Doviak, Dr. Robert Weinbeck, Dr. Ira W. Geer, Dr. James R. Fleming, Dr. Janusz Borkowski, Dr. Raymond L. Lee, Jr., and to Dr. Joseph M. Moran, for reading the manuscript and providing valuable comments. Many thanks are addressed to Mrs. Betty Anderson for her technical assistance. A very special appreciation is extended to Mrs. Annabelle Sherba and Mr. Charles Sherba for proofreading and editing the text.

Zbigniew Sorbjan

Norman, Oklahoma
Fall, 1996

There is little hope that he who does not begin at the beginning of knowledge will ever arrive at its end.
(H. von Helmholtz)

CHAPTER ONE

BASICS

Meteorology, Development of Atmospheric Sciences, Ancient Experiments, Meteorology and History, Meteorology and the Arts.

METEOROLOGY. Interest in weather phenomena is almost as old as civilization. For instance, the word *meteorology* was invented by the Greeks over two thousand years ago. In the fourth century B.C., the Greek philosopher Aristotle (384-322 B.C.) published a treatise on the subject of weather, entitled *Meteorologica*. The word *meteor* comes from two Greek words *meta* - beyond, and *eora* - suspension. With this word, Aristotle meant atmospheric elements such as rain, snow, hail, wind, thunder or lightning, and also earthquakes, comets and Milky Way. The word *meteor*, presently used by meteorologists, is sometimes confused with the words *meteor, meteoroid* or *meteorite*, which do have the same roots, but in astronomy indicate extraterrestrial objects entering the atmosphere.

Aristotle's *Meteorologica* consisted of four books. The first book described comets, winds, rains, clouds, rivers, springs, dew, hail, and climate. It also explained the theory of four elements: fire, water, earth and air. The second book discussed seas, earthquakes, lightning, and thunder. The third book examined hurricanes, whirlpools and light. The fourth book dealt with properties of hot, cold, wet and dry objects.

The word *atmosphere* is also derived from the Greek word, *atmos* which means vapor. For ancient Greeks the atmosphere meant a "region of vapor".

Aristotle (384-322 B.C.)

Aristotle, the greatest of all Greek philosopher-scientists was born in Stagira (Macedonia), as the son of a physician to the royal court. At the age of 17, Aristotle entered Plato's Academy in Athens. He remained there for 20 years, first as a student and later as a teacher. He allegedly lisped, had small eyes and thin legs. In 343 B.C., Aristotle became the tutor of Alexander the Great in Pella, the Macedonian capital. In 335, he returned to Athens and established his own school, the Lyceum. Aristotle systematically recorded the knowledge of his time. Aristotle's lectures were collected into nearly 150 volumes; only about one-third survived. Those which survived were adopted in the twelfth and thirteenth centuries by Christian Europe through translations from Arabic. Aristotle's explanations were speculative and mostly wrong from the point of view of modern science, but marked the beginning of scientific research. His encyclopedic works served as the accepted basis in science until the time of Nicolaus Copernicus and Galileo Galilei, who laid the foundation of modern science.

DEVELOPMENT OF ATMOSPHERIC SCIENCES. The history of meteorology has three general periods: from ancient times to 1500 A.D., from 1500 A.D. to 1800 A.D., and from 1800 A.D. through modern times.

In the first period, circa 600 B.C. to 1500 A.D., meteorology was not an independent discipline, but a part of general science. It was mostly based on pseudo-scientific speculations.

Scientific discoveries during this period were contributed by various peoples of the ancient world such as the Babylonians, Sumerians, Chinese, Hindus, Egyptians, Greeks, Romans, Arabs and others. For instance, the Babylonians laid the foundations of early mathematics. They knew the four cardinal directions: south, north, east and west, and also the intermediary directions, which they called by names made up of the four cardinal words. The Egyptians defined weights and measures, invented ingenious water-clocks, and introduced the 365 day year. The Chinese invented the compass, made astronomical observations and meteorological predictions.

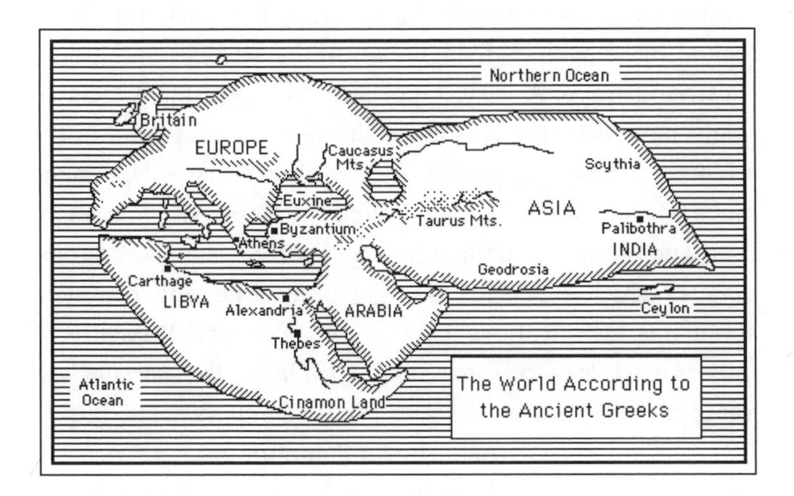

The most methodical contribution to science was made by the Greeks who developed geometry, logic, and philosophy. They performed meteorological observations and, in turn, created physical theories. For instance, about 400 B.C. Hippocrates wrote *Airs, Waters and Places,* a study of climate and medicine. A few years later, Aristotle published his *Meteorologica.* About 300 B.C., the Greek philosopher Theophrastus, wrote a meteorological treatise *On the Signs of Rain, Winds, Storms and Fair Weather.*

A Grecian vase

The next great civilization of the Romans did not add significantly to science. Roman contributions lay mainly in the

field of organization, law, medicine, agriculture, building roads and aqueducts. One of the Roman philosophers interested in meteorology was Seneca (2-65 A.D.). In his *Natural Questions,* Seneca discussed various atmospheric phenomena. In the first three books he examined theories of the rainbow, halo, thunder, lightning and wind. Book Four dealt with clouds, rain, dew, hail, frost, snow and ice (the book ended with Seneca's moralizing against the luxury of cooling drinks with ice). Seneca's sources for his work included mostly Greek writers.

The period of European history which began with the fall of the Roman Empire (the western Empire fell in the 5th century) and ended with the beginning of the Renaissance (15th century), was known as the *Middle Ages*. This period has been looked upon as one of stagnation in the sciences and arts in Western Europe. A very important but unfortunate incident might have been the reason for the slow development of the sciences during that period. In the 4th century A.D., fanatic mobs burned down the Library and the University in Alexandria. The Library contained hundreds of thousands of scrolls, many of which were original single copies. One could comment about this historical event by quoting Bertrand Russell who said: "*Knowledge is very much more often useful than harmful, and fear of knowledge is very much more often harmful than useful*".

About the turn of the first millennium A.D., science was carried on by the Arabs who were known for their achievements in mathematics, optics and astronomy. Due to territorial expansion, the Arabs became acquainted with the achievements of the Greeks and other cultures, especially those of the Far East. These achievements were later passed on to Europe. For example, the discovery of paper (in A.D. 704), was transferred from China, and the decimal system from India (thanks to the work of Al-Khwarazmi of Baghdad). The Arabs learned from Hindu scholars the use of the zero sign, a great mathematical discovery (who would have thought that so little could mean so much?). They coined the term *algebra*. The notation for numbers used in mathematics is still called *Arabic*. The Arabic

Medieval Scholar

numbers were introduced in Europe in 1202 by Leonardo (Fibonacci) of Pisa. Aristotle's *Meteorologica* was translated from Arabic to Latin by the Italian, Gerard of Cremona, in about the year 1170, and first printed in Padua in 1474.

In medieval Europe, research was chiefly performed in libraries and based on the authority of ancient authors. Science was replaced by scholasticism, based on the dogmatic authority of the scriptures. Before the year 1250, the Church forbade the reading of Aristotle's natural philosophy, but later his teachings were reinterpreted by Thomas Aquinas (1225-1274). As a result, Aristotle's works were accepted as containing all possible knowledge and, therefore, were adopted without question.

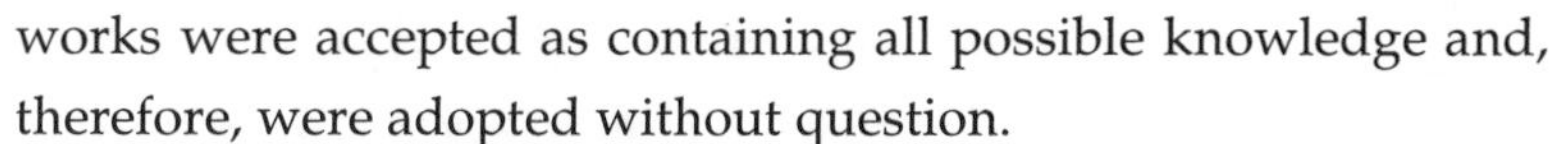

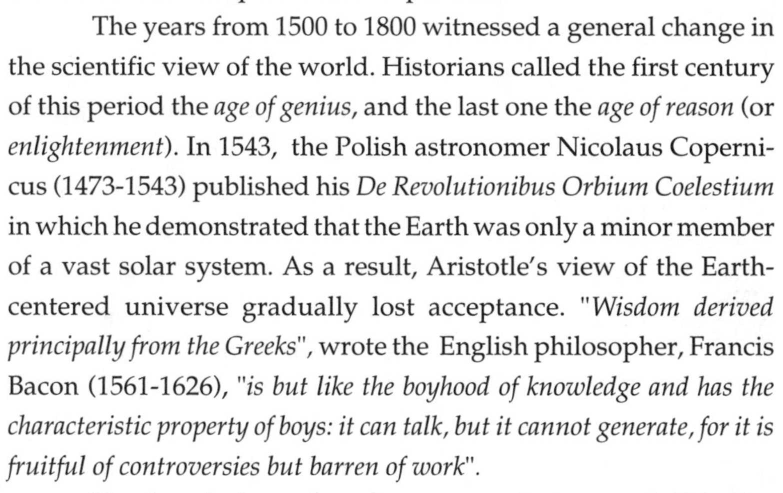

The years from 1500 to 1800 witnessed a general change in the scientific view of the world. Historians called the first century of this period the *age of genius*, and the last one the *age of reason* (or *enlightenment*). In 1543, the Polish astronomer Nicolaus Copernicus (1473-1543) published his *De Revolutionibus Orbium Coelestium* in which he demonstrated that the Earth was only a minor member of a vast solar system. As a result, Aristotle's view of the Earth-centered universe gradually lost acceptance. "*Wisdom derived principally from the Greeks*", wrote the English philosopher, Francis Bacon (1561-1626), "*is but like the boyhood of knowledge and has the characteristic property of boys: it can talk, but it cannot generate, for it is fruitful of controversies but barren of work*".

The foundations of modern meteorology were laid in Europe with the invention of meteorological instruments and the introduction of meteorological observations. Men no longer applied mysticism to explain weather. In 1593, Galileo Galilei invented the *gas* thermometer. Fifty years later, in 1644, Evangelista Torricelli invented the *mercury* barometer. In 1661, Robert Boyle and Edme Mariotte formulated the first law on the nature of gases. In 1687, Isaac Newton formulated the laws of mechanics. Equations of fluid motion were derived by Leonhard Euler in 1752. In the

1780s, the foundation of modern chemistry was laid by Antoine Lavoisier. "*Perhaps the men of genius are the only true men*", wrote Aldous Huxley. "*In all the history of the race, there have been only a few thousand real men. And the rest of us - what are we? Teachable animals. Without the help of the real men, we should have found out almost nothing at all*".

In the past two centuries meteorology experienced a rapid expansion of its scientific horizons, influenced by developments in the related disciplines such as mathematics, physics, chemistry, and electronics. For instance, the first mathematical solution of equations describing a simple atmospheric motion was obtained by Valfrid Ekman in 1905. *Synoptic meteorology* was developed by Vilhelm Bjerknes and his followers after World War I. The first computerized weather prediction was accomplished in 1950. The first investigations of the vertical structure of the atmosphere were performed by using kites, balloons, and towers. Currently meteorologists have at their disposal the most modern technology available, such as *sodars* (sound detectors of the atmosphere), *lidars* (light detectors), *radars* (radio-wave detectors), *airplanes* and *satellites*. They also employ the largest and the fastest electronic computers, which are capable of adding hundreds of millions of numbers in seconds.

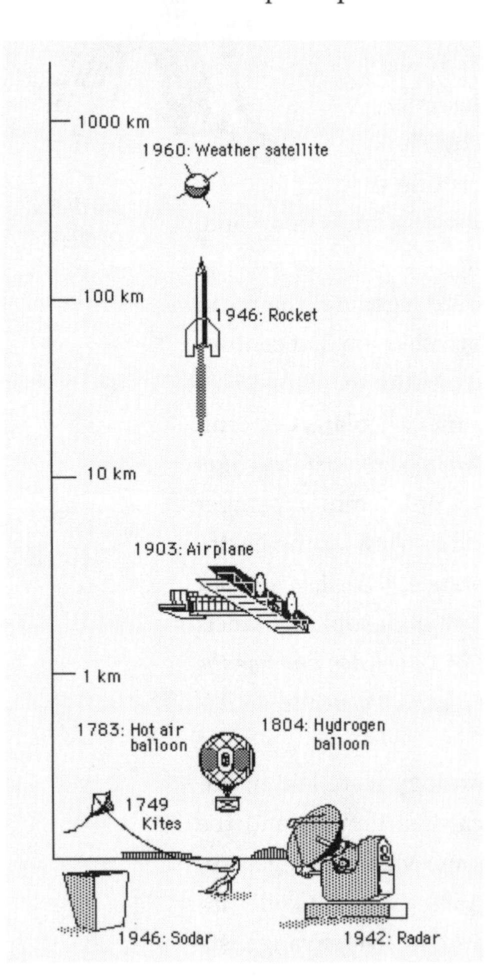

Exploration of the atmosphere

ANCIENT EXPERIMENTERS. The urge to ask the questions *why* and *how* about the world has always been an attribute of inquisitive minds. On many occasions the answers to such questions were derived by performing simple experiments, many of which are described in this book.

The ancient Greeks performed simple experiments in their investigations of the world. As a result, they were repeatedly able to achieve astonishing results, such as the one described below.

Sometime in 330 B.C, the Greek philosopher, Eratosthenes, learned from a traveler that in Syene (Aswan in Egipt, now the site of a huge dam), on a midsummer's day (summer solstice), at noon on June 22, the Sun was exactly overhead. Wishing to make a similar measurement in Alexandria, he used a vertical rod placed in the bottom of a hemispherical bowl. On a midsummer's day in Alexandria, the rod cast a shadow, the length of which measured 1/50 of the sphere's circumference. Eratosthenes felt that the result was worth explaining. After some consideration, he came to the correct conclusion that this difference in angles between the parallel beams of sunlight and the local verticals in Syene and Alexandria (marked as S and A in the figure below) was caused by the curvature of the Earth.

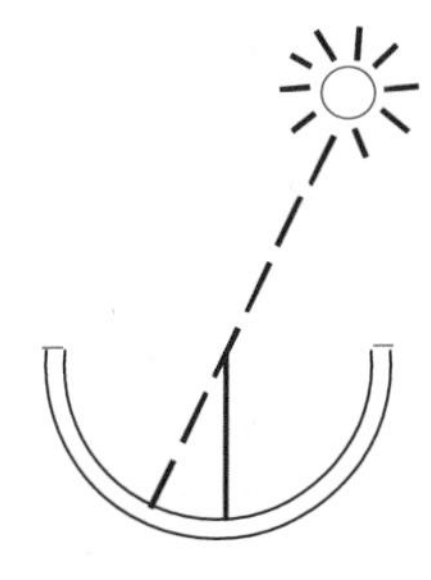

Shadow in a bowl

He knew that Syene was located 5000 stades (Greek units of length) south of Alexandria. Unfortunately, no one today is sure of the exact length of 5000 stades. Probably 1 stadium is about 1/6 km, because the distance between Alexandria and Syene is about 800 km. Consequently, Eratosthenes calculated that if 1/50 of the Earth's circumference was equivalent to 5000 stades, then the Earth's circumference was: 5000 stades x 50 = 250,000 stades = 41,666 km. The modern value is 40,000 kilometers. Eratosthenes' result was remarkably accurate. Nevertheless, that result seemed too large to the

Eratosthenes' calculation

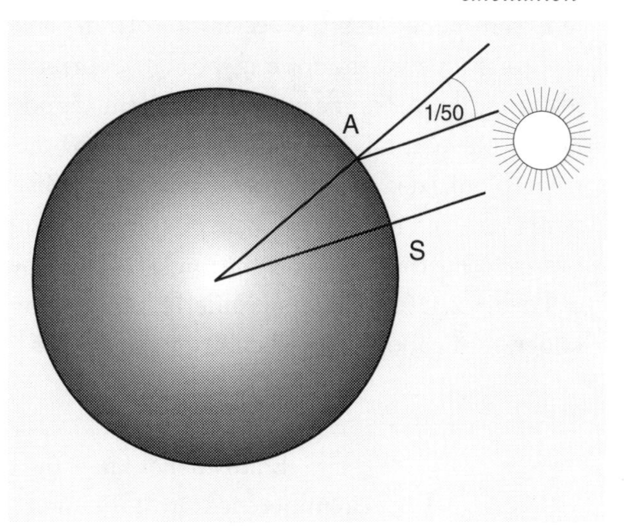

ancients and was not verified until the voyage of the Portuguese explorer, Ferdinand Magellan, in 1522.

Measuring units. There have been many different measuring units used throughout history. Stadium, the unit used by Eratosthenes is but one example. The first systems of measures and weights were developed in ancient Egypt and Babylon around 3500 B.C. Some elements of the old measuring systems are still used; for example, units of time. Claudius Ptolemy (100-160), the celebrated Greek physicist, who lived in Alexandria, adopted the Babylonian sexagesimal system and divided each hour into 60 minutes, and each minute into 60 seconds. The Latin translation of the names of these units of time were rendered as *partes minutae primae* and *partes minutae secundae*. These were the sources for the words *minute* and *second*. In modern science the metric system (*decimal* system with number 10 as the basis) is used. This system, called SI, from the French words, *Systeme International*, was first adopted during the French Revolution in 1799, to standardize the variety of local measurements. Our book also incorporates the SI system, which uses *meter* (= 100 cm = 1000 mm) as the unit of length, *second* as the unit of time, *kilogram* (= 1000 g) as the unit of mass, *Kelvin or Celsius degrees* as the unit of temperature. Originally, the standard length of one meter was defined as one ten-millionth the distance between the North Pole and the Equator, measured along the meridian passing through Paris. One kilogram was the mass of one liter of liquid water at its maximum density at 4°C. Thomas Jefferson (the author of the Declaration of Independence) and Alexander Hamilton were instrumental in adopting the metric currency system for the U.S., but failed to change the metric measurement system. Opposition to using the system of the French Revolutionaries was too strong, and the British system has remained from colonial days. Consequently, in the U.S. distance is still measured in miles, yards, feet, and inches; weight in pounds and ounces; volume in gallons, quarts, pints, etc. For those who use the English system, note that 1 meter = 39.37 inches, or 1 inch = 2.54 centimeters. Similarly, 1 kilogram = 2.2 pounds, or 1 pound = 0.45 kilogram, 1 gallon = 0.0038 cubic meter = 3.785 liters.

Eratosthenes knew the value of the ratio π of the circle's circumference C to its diameter D. In fact, Egyptian priests and architects knew the value of π as early as 1800 B.C. According to their formula from the Rhind Papyrus in the British Museum, $\pi = (16/9)^2$, which is equal to 3.16. Knowing that $C = \pi D$, Eratosthenes could have obtained the Earth's radius with a relative error of less than 5% (our result is 6370 km). It hardly seems possible to obtain such a brilliant result merely from a stick of wood.

Connections. Note the connections among the individuals described on the pages of this book. The names of Euclid, Ptolemy I, Aristotle, and Eratosthenes have already been mentioned. For example, one of Aristotle's pupils was Alexander the Great, the Macedonian conqueror and the founder of the city of Alexandria (in 332 B.C.), situated at the mouth of the Nile. His ablest general became the King of Egypt, as Ptolemy I (ruled c. 285-246 B.C.), and was also the originator of a dynasty which later included Queen Cleopatra (ruled in 51-30 B.C.). Ptolemy I fostered the advancement of science in his kingdom. Inspired by the great library he had seen in Babylon, he established the Library of Alexandria. The Library grew to be the largest of the ancient world, containing upward of 100,000 papyrus scrolls. Ptolemy also founded a university which was dedicated to the *Muses* (Goddesses of the Arts) and called, therefore, *Museum.* The university allegedly had 14,000 graduates during the few centuries of its existence. Its first professor of geometry was Euclid (315-250 B.C.). Eratosthenes of Cyrene (275-195 B.C.) was the chief librarian at the Museum of Alexandria.

METEOROLOGY AND HISTORY. In the past, as well as in the present, weather played an important role in influencing human life. Thales of Miletus (620-540 B.C), the founder of Greek philosophy, was the first person known to apply in practice the knowledge of meteorology. By carefully observing weather trends, he predicted an abundant olive crop one particular summer. He wisely bought all of the olive presses in his immediate vicinity. Sure enough, his prediction came true and he became a rich man. As a result of his new found wealth, Thales was able to devote the rest of his life to philosophy. The author hopes that the readers of this book will benefit no less from their understanding of the weather.

Typhoon waves

Every change in the conditions of the atmosphere brings joy to some and sorrow to others. On the great scene of history the destiny of nations has often been depended on some unexpected vicissitude of the weather. For instance, in June 1281 A.D., the Mongol Emperor Kubilai Khan (1260-1294), the Conqueror of China, and grandson of Genghis Khan, launched an invasion against Japan with 140,000 troops on 4400 ships. Because his army was so powerful, Kubilai Khan would most likely have succeeded if a sudden weather change had not occurred. Without warning, the whirling winds of a typhoon ("big wind" in Chinese) rose from the sea. The storm tore at the Khan's warships and crushed them against the shore. As a result of this cataclysm, more than half of the Mongolian warriors died. The event was described by Marco Polo (1254-1324), an Italian traveler who was visiting the court of Kubilai Khan at that time. Marco Polo reported that some warriors drowned in the sea while others made it ashore, only to be slaughtered by Japanese defenders. The storm destroyed the Mongolian army, but created a national Japanese myth about the special guardianship of *Kamikaze*, which means "divine winds".

Armada

In 1588, King Philip II of Spain (1527-1598) deployed an Armada of 130 warships carrying 27,000 men in order to attack England, then governed by Queen Elizabeth (1533-1603). The history of England would have been very different if, again, the weather had not intervened. Unfortunately for the Spaniards, a violent five-day storm crushed many of their ships against the rocky coast of Scotland. In an attempt to escape, the remaining ships fled westward, but were hit near Ireland by yet another fierce storm. The calamity destroyed the

Spanish navy with only half of the fleet returning.

The reign of the Emperor of France, Napoleon Bonaparte, ended due to bad weather. His military strength was initially softened during the severe winter of 1812 in Russia, and then finally crushed during a rainy campaign which took place in June 15-18, 1815. The decisive battle of Napoleon's last campaign occurred near the village of Waterloo (Belgium) on the 18th of June. The battle involved the French army against Anglo-Dutch forces commanded by the Duke of Wellington at first, and later also the Prussian army under Prince Blücher. On the 17th, thunderstorms occurred early in the afternoon while the French were in the process of attacking the Anglo-Dutch force. Rains turned the ground into swamp. Consequently, the French advance was greatly slowed down. This allowed Wellington's lighter force to be almost completely preserved and withdrawn to a better position. The rains during the night from the 17th to the 18th saturated the ground even more, so that the battle could not start before noon. For the first few hours the battle went in favor of the French. The arrival of fresh Prussian soldiers around 4 p.m. turned the tide of fighting and led to Napoleon's defeat.

Bonaparte

METEOROLOGY AND THE ARTS. Meteorology, the science of the atmosphere, weather and climate, touches our daily lives in significant ways and is inextricably connected to the humanities: philosophy, literature, and the fine arts.

In music, for instance, atmospheric conditions have also provided composers and lyricists with unrestrained sources of inspiration, not limited to a specific genre. This fact can be supported by an endless number of examples. Vivaldi's classical *Four Seasons* hauntingly paints the musical contrasts of the different seasons of the year. Equally illustrative are the stories of thunderstorms, beautifully unleashed in the music of Beethoven's *6th Symphony* and in Rossini's *The Barber of Seville.*

Painters have always been sensitive to atmospheric phenomena. They have carefully studied the sunlight, the color of the sky, clouds, sunrises, sunsets, and rainbows. For instance, Leonardo da Vinci in his *Treatise on Painting* noted that "*objects seen through a fog seem to be further away from the observer, and therefore appear to be larger than they really are*".

Inspired by the beauty of nature, painters have provided us with memorable scenes of nature generously instilled with blue skies and vibrant sunsets. "*I was walking along the road*", wrote Edward Munch, a famous Norwegian painter. "*The sun set. I felt a tinge of melancholy. Suddenly, the sky became a bloody red. I stopped, leaned against the railing, dead tired and looked at the flaming clouds that hung like blood and a sword over the blue-black fjord and city. I stood there trembling with fright. And I felt a loud, unending scream piercing nature; it seemed to me that I could hear the scream. I painted this picture, painted the clouds as real blood.*"

Paintings of the Italian school have been easily distinguishable due to the characteristic blueness of the sky. Artists of central and northern Europe used softer and more subtle hues. Their coral pink and willow green, half-tone landscapes under pale blue or gray misty skies, were filled with diffused sunlight which did not cast shadows. On the other hand, American western landscapes were painted in complementary colors of blue, green, red and orange.

Literature is yet another field steeped with references to the weather. In Shakespeare's "Romeo and Juliet", the entire plot hinged upon the influence of a hot day. The fatal fight, in which Tybalt is slain, is evoked by the impact of the temperature upon the principal characters of the play. Benvolio realizes that effect, and in an attempt to restrain his vigorous companion, says: "*I pray thee, good Mercutio, let's retire. The day is hot, the Capulets abroad, and if we meet, we shall not' scape a brawl. For now, these hot days, is the mad blood stirring*".

Joseph Conrad (Korzeniowski) in *Typhoon* vividly described the horror of severe weather: "*They were running all over the sea trying to get behind the weather*". In contrast, Homer in the *Odyssey*, described the ideal weather on the top of Mount Olympus as never being shaken by wind, rain or snow, which the gods enjoyed.

Science which is not rooted in experiment is empty and full of errors.
(Leonardo da Vinci)

CHAPTER TWO

AIR

Theories about Air, Volume and Weight of Air, Density of Air, Height of the Atmosphere, Alchemy of Air, "Fixed Air" - Carbon Dioxide, "Inflammable Air"-Hydrogen, "Noxious Air"- Nitrogen, "De--phlogisticated Air"- Oxygen, Other Atmospheric Gases, Gas Molecules, Abundance of Atmospheric Gases, Structure of the Atmosphere, Origin of the Atmosphere, Past Climates.

THEORIES ABOUT AIR. What is air? Is it anything at all? Does it occupy space? Does it have weight? The answers to these questions are quite obvious now. But, the correct answers have been known only for the last 250 years. Before that time, only primitive speculations and beliefs existed on the subject.

The Greeks were the first to recognize that air had substance. Anaximenes (550-475 B.C) thought that air was a fundamental element of the universe and could become water when cooled, or earth when compressed. He speculated that air contained an essence he called *pneuma*, and believed that this essence supported the universe in the same way that air supported human existence. He also believed that air became cooler with height as a result of a progressive decrease in the intensity of sunlight reflected from the surface of the Earth. Anaximenes maintained that beyond a certain point, the Earth's atmosphere was made of a fire-like substance, called *ether*, which was responsible for thunder and lightning.

VOLUME AND WEIGHT OF AIR. In the fourth century B.C., Aristotle proved experimentally that air occupies volume (most likely by using a procedure described in Experiment 2.1). He also expected air to have weight. To test his hypothesis, he weighed an empty animal bladder and then weighed the same bladder when it was filled with air. Because his measurement was not very accurate, he was surprised to find that both weighed the same. Consequently, Aristotle concluded that air had no weight and that air's lightness was permanent.

2.1. AIR OCCUPIES VOLUME

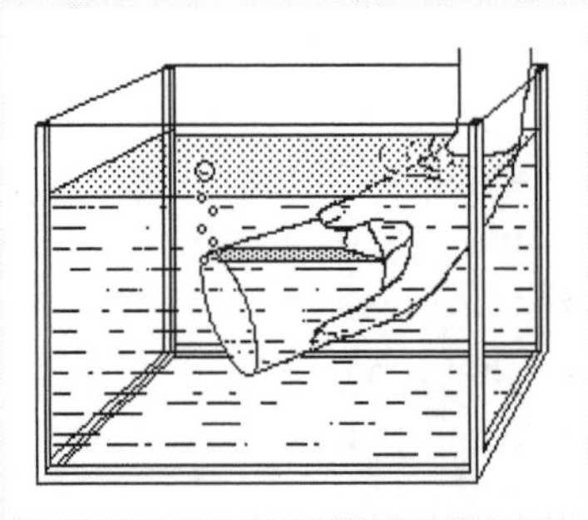

Materials: aquarium or large water bowl, drinking glass, water.

Procedure: Fill an aquarium with water to about three-quarters capacity. Holding a drinking glass mouth downward, submerge it into the aquarium. Water will not enter the glass full of air. Tilt the glass and let some of the air escape. Bubbles will rise up to water surface from glass indicating a presence of air.

For centuries after Aristotle, people believed that air was weightless. In 1638, Galileo Galilei, a famous Italian philosopher and mathematician, devised an experiment to determine if air had measurable substance (weight). He fitted a large, narrow-necked glass flask with a valve and forced a great quantity of air into it with a syringe. He carefully weighed the flask, released the compressed air, and weighed the flask again. From this experiment, he concluded that air did indeed have weight.

The idea that air had weight contradicted common sense even a few decades after Galileo's experiment. When King Charles II of England in 1664 visited the Royal Society, to witness an experiment which attempted to weigh air, the event provoked silly laughter. The ladies and gentlemen of the Royal Court found this experiment to be very peculiar because: "*anyone can see that there is nothing in air to be weighed*".

The royal visit

The "U.S. Standard Atmosphere" is a document published by the U.S. Government, which describes an idealized atmospheric state as a function of height. According to this document, 1 m^3 of air weighs 1.225 kg near the Earth's surface. This means that the air contained in a one-liter bottle weighs 1.225 grams (0.0429 ounces).

Galileo Galilei (1564-1642)

Galileo Galilei was born on February 15, 1564 in Pisa, Italy. He studied and later taught at the University of Pisa. He was a short, energetic man, with reddish hair. For his bold statements, he was disliked by some of his university colleagues in Pisa. In 1592, he moved to Padua, where he remained for 18 years. During his residency at Padua, he took a Venetian mistress named Maria Gamba, by whom he had 2 daughters and a son. In 1610, Galileo moved to Florence, where he took the position of a philosopher in the Court of the Grand Duke of Tuscany, Ferdinand II de Medici. Galileo's main work, *The Dialogue Concerning the Two Chief Systems of the World, the Ptolemaic and the Copernican*, was published in 1632. It was written in the form of dialogues of the two supporters of his ideas, Sagredo and Salvati, and a supporter of the Aristotelian point of view, Simplicius. *The Dialogue* was written with the permission of the newly elected Pope Urban VIII, whom Galileo knew personally, and who revoked the 1616 decree forbidding Galileo to write about the Copernican system. Nevertheless, in 1633, Galileo was charged and tried by the Roman Inquisition for teaching Copernican ideas. The trial involved the highest theological authority of the Catholic Church in a scientific dispute with the most prominent scientist of the day. Galileo was forced to publicly recant his ideas as being false and opposed to the Holy Bible. Threatened with torture, Galileo agreed to comply with that order. However, as he rose to his feet after abjuring on his knees before the Cardinals, he allegedly muttered: "epur si muove", which means "even so, it moves". Galileo's life was spared but he was placed under house arrest for the rest of his life. Blind for the last four years of his life, he died in Florence in 1642. The Grand Duke Ferdinand II wanted to erect a suitable tomb for Galileo, but was warned not to do anything which might trigger an unfavorable response from the Holy Office.

It took the Vatican centuries to admit its errors. *The Dialogue* was removed from the list of publications banned by the Church in 1757. Galileo was finally exonerated by Pope John Paul II in 1992, 359 years after the trial.

2.2. AIR HAS WEIGHT

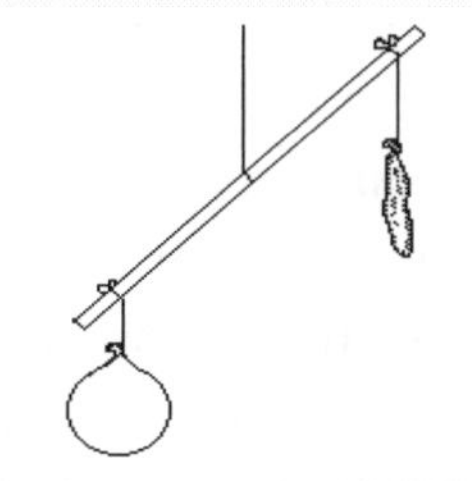

Materials: two balloons, wooden rod, string.

Procedure: Inflate two large balloons to the same size. Tie them at each end of a wooden rod. Suspend the rod by a string and adjust it so the rod balances. The air in the room should remain still. Puncture one of the balloons with a needle. The side of the rod with the inflated balloon will drop.

Explanation: The rod with the inflated balloon drops because the balloon contains compressed air. Notice that this experiment only indirectly proves that air has weight and solely shows that the compressed air in the balloon has greater weight than the same volume of atmospheric air.

DENSITY OF AIR. The ratio of mass and volume is called *density*. The concept of density is quite old. In 1025 A.D., Al-Biruni, an Arab physicist living in Persia, invented the "*overflow-can* " method of finding the density of a body. More correctly, Al-Biruni introduced the idea of *specific gravity*, which is the ratio between the weight of a body and an equal volume of water. Regardless of what unit is used for weight, the ratio remains the same for any given substance. The ratio *s* :

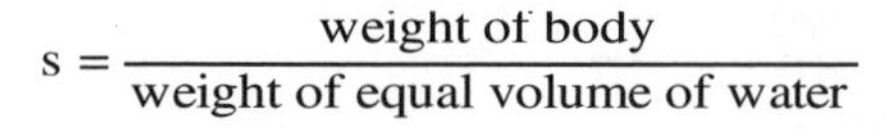

$$s = \frac{\text{weight of body}}{\text{weight of equal volume of water}}$$

shows how many times the density of the substance is greater over that of water. It can be noted that in the metric system, the density of water is 1 g/cm^3. Consequently, in that system, the specific gravity of the substance and its density are numerically equal.

Overflow can

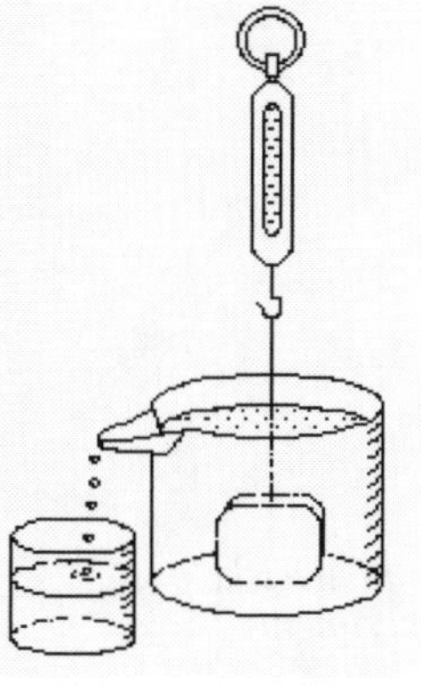

Heat affects air density. The density of atmospheric air decreases when temperature increases. This phenomenon has an impact on aviation. In very hot weather, atmospheric air becomes very thin and takeoff distances have to be significantly extended. If runways were too short, large airplanes would be unable to depart under such conditions. An example of this occurred several summers ago at the Sky Harbor Airport in Phoenix, Arizona; the runway was too short for most aircraft when temperatures reached 120 F (45°C).

Unfortunately, Al-Biruni's method cannot be applied to find the density of air. Air is lighter than water. When a bottle is filled with air and immersed in water, it will float on the surface. Therefore, how can the density of air be found?

After hearing of Galileo's work, Otto von Guericke (1602-1686) performed his own experiments. In 1654, he removed air from a glass vessel with an extraction air-pump which he had invented. As a result, he established that the vessel weighed less after the air was extracted. By comparing the weights of an empty vessel, and the vessel filled with air, and then filled with water, he calculated that the air's specific gravity $s = 1/947$, or that air was 947 times less dense than water. Since the density of water is 1000 kg/m^3, the resulting density of air equaled 1.05 kg/m^3. Today, the accepted standard value (i.e., at sea level, and at the temperature of melting ice) of the air's density is 1.29 kg/m^3.

Von Guericke's experiment

Otto von Guericke (1602-1686)

Otto von Guericke was born on November 12, 1602 in Germany. He attended the University of Leipzig (1617-1620), the University of Helmsted (1620), studied law in Jena (1621-1622), and law, mathematics and engineering in Leiden. In 1626, he became an alderman of the city of Magdeburg. From 1618 to 1648, Germany went through a violent period in its history, called the Thirty Years' War. This war between Protestants and Catholics was a result of a dispute over whether a Roman Catholic prince or a Protestant one should become King of Bohemia. Although the combatants were non-German (Sweden, Denmark, Spain and France), the war was fought mostly on German soil. In 1631, as a supporter of the Protestants, von Guericke became Quartermaster General to the Swedish King, Gustav Adolf I. In 1635, he performed similar duties for the Elector of Saxony. In 1646, von Guericke became the mayor of Magdeburg.

HEIGHT OF THE ATMOSPHERE. The height of the atmosphere was first evaluated in the year 1025 by an Arabian scientist, Alhazen (965-1039). Alhazen understood that twilight was caused by the reflection of sunlight from particles in the atmosphere. He measured the duration of twilight and obtained an average value of about 36 minutes (= 0.6 of an hour). During this period, the Earth turns through: 0.6 h / 24 h x 360°=9°. In the accompanying Figure: R / (R + H) = cos 9°, where R is the radius of the Earth (= 6115 km, which is Erastothenes' result expressed in kilometers), and H is the

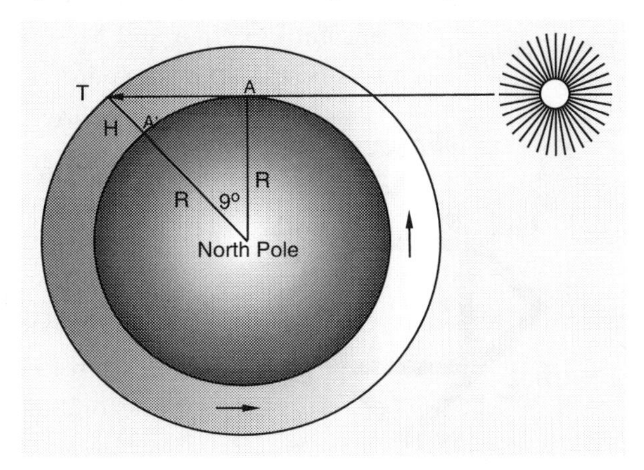

Alhasan's calculation

height of the atmosphere. Using a calculator one can obtain that cos 9° =0.988. This yields:

$$H = \frac{R\,(1 - \cos 9^{o})}{\cos 9^{o}} = 6115\,\frac{1 - 0.988}{0.988}\,\text{km} = 74.3\ \text{km}$$

Presently, it is known that there is no distinct top of the atmosphere. Since air density gradually decreases with height, one might define the top of the atmosphere at that level where air density is very small, for instance, 10,000 (quite an arbitrary value) times smaller than it is on the surface of the Earth. That level is about 70 km above the surface, indicating that Alhazen's estimate made sense.

ALCHEMY OF AIR. The science of the nature and composition of material substances is called *chemistry*. The early and primitive form of chemistry was called *alchemy.* The word *alchemy* is derived from the Arabic, *alkima,* in which "*al*" is the definite article and *kima* is believed to come either from the Greek, *chyma,* meaning "cast a metal", or from *chem,* "the dark land", the ancient Egyptians' name for their country. *Alchemy* was probably initiated in Hellenist Egypt. However, various chemical elements were discovered prior to the Greeks. For example, copper was known to the Egyptians as early as 3000 B.C.; bronze (a mixture of copper and tin) was known to the ancient Egyptian and Mesopotamian civilizations; and, iron was produced by the Hittites about 2000 B.C.

The four elements of Aristotle

Aristotle believed that the entire terrestrial region was composed of four elements: *earth, water, air,* and *fire,* built in consecutive shells. All things on the Earth were supposedly derived from these four elements. Fire and water were in opposition to each other, while water and air combined with each other. Aristotle also invented another element, *quintessence* (meaning "the fifth element"), which was generally reserved for the heavenly spheres.

The four elements were associated with four basic features: *heat*, *cold*, *moisture* and *dryness*. The elements could change one into another, and each was potentially latent in the others. The transforming agent was the Sun. The Sun could draw a moist and cool substance from the water and a hot and dry substance from the Earth. This dry exhalation formed "the origin and natural substance of the winds", and was also responsible for earthquakes and comets. In combination, the cool and moist substances formed air. Air could be turned into clouds, rain, snow, frost or dew.

Another Greek scientist, Anaxagoras (488-428 B.C.), did not believe that the structure of matter was based on four elements. In his view, matter did not change regardless of how small it became. For instance, a piece of rock crushed to powder still remained a rock. Similarly, Leucippus of Miletus (about 440 B.C.), and his student Democritus (460-370 B.C.), both thought that *matter* was composed of very small, indivisible particles, called *atoms* (from the Greek word *atoma* which means "indivisible"). Democritus held that atoms were in constant motion but differed from each other in shape and arrangement.

After the Muslim conquest of Egypt in 640 A.D, the Greek alchemic treatises were translated into Arabic. The introduction of alchemic knowledge and techniques into western Europe occurred in the 12th - 13th centuries through Spain, where the first Latin translations were prepared.

The idea of transforming one kind of matter into another was the most important problem of alchemy practiced in Europe. The main purpose was to obtain the "perfect metal" (gold) from less valuable compounds or the universal remedy "panaceum". Although the goal was never achieved, many other important discoveries resulted from this investigation. For instance, about 1709 the German alchemist Johan Friedrich Böttger, was hired by the Elector of Saxony to find a formula for making gold. Instead of gold, Böttger discovered a method for making white porcelain, which at that time was even more precious than gold.

Francis Bacon, the 16th Century English philosopher, wrote: "*Alchemy may be compared to the man who told his sons that he had left them gold buried somewhere in his vineyard. The sons by digging found no gold but by turning up the mold about the roots of the vines, procured a plentiful vintage*".

Alchemist at work

Nineteen centuries after Aristotle, alchemy was still the subject of speculative investigation. René Descartes (1596-1650) in his treatise *Discourse on Method*, expressed a belief that air (and also all other earthy substances) was composed of a single fundamental material. Particles of this basic material, varied in size and in shape. The elementary particles of air were "*layered upon one another without being interlaced*". Fundamental elements of solids were "*hooked and bound to each other*". Liquids were made of smooth, slippery particles and therefore did not bond. Descartes' explanation generated great interest in Europe and many scientists became occupied in finding this mysterious basic element.

René Descartes (1596-1650)

René Descartes (a Latinized name: Renatus Cartesius) was born in La Haye, France, on 31 March 1596. After receiving a law degree from the University of Pointiers in 1618, at the beginning of the Thirty Years' War, he joined the army as a gentleman volunteer for Prince Maurice of Nassau. Later that same year, he was assigned to the Duke of Bavaria's army on the Danube. He was not participating in actual warfare. Instead he worked out his (Cartesian) philosophy and geometry. On 10 November, 1619, allegedly after a mystic revelation, Descartes reached a radical conclusion. He realized that in order to discover knowledge, he had to carry out the whole program by himself, methodically, by doubting everything except his existence (his famous "*Cogito ergo sum*", "I think, therefore I am"). In 1649, Descartes traveled to Stockholm where he was offered a position as a court philosopher to the Swedish Queen Christina. He died of pneumonia during his first winter in Stockholm on 11 February 1650. His body, all but the head, was returned to France. His skull remained in Sweden until 1809.

Georg Ernest Stahl (1660-1734), a professor of medicine and chemistry at Halle (Germany), introduced the idea that combustible objects contained *phlogiston* (from the Greek word *flogistikos*, "something which burns"). This was a hypothetical material which supposedly was released into the air to extinguish a flame, whenever a substance was burned or rusted. For instance, phlogiston could be transferred from charcoal (rich in it) to a metal ore (poor in it) in a process in which charcoal burned and the ore was converted into metal. Products of combustion were without phlogiston and thus could not burn. Because rusting substances gain weight and burning substances appear to lose weight, the weight of phlogiston was regarded to be either positive or negative.

An officer and a gentleman. Descartes was a contemporary of Cyrano de Bergerac, and lived in the Paris of Cardinal Richelieu, so vividly described by Alexandre Dumas (1824-1895) in *The Three Musketeers*. Once when an inebriated young man offended Descartes' lady companion, the philosopher reacted in the fashion of Cyrano. In the fracas that followed, Descartes flicked the sword out of the drunkard's hand. Then, he generously spared the lout's life, because he felt that the rascal was too filthy to be slaughtered before a lady.

In the beginning of the eighteenth century there were speculations rejecting the doctrine of four elements in favor of only two: water and fire. It was believed that water could be changed into earth (e.g., evaporated water always left behind a solid residue), that air was a result of a combination of water and fire (e.g., water could be evaporated into a vacuum by heating), and that air was "fixed" (absorbed) in a variety of animal, vegetable and mineral substances.

Gas. The word *gas* was first used in 1632 by the Flemish chemist, Johan Baptista van Helmont (1580-1644), in his book *Ortus Medicinae* (at this time gases were called "airs"). During one of his experiments he noticed that even though 62 pounds of charcoal burned into one pound of ash in a closed vessel, the vessel's weight did not change. He concluded that the other 61 pounds of charcoal had turned into "wild spirit". He named it *gas*, a word derived from the Greek word *chaos* (note that in Flemish "ch" is pronounced as " g"). Van Helmont died in 1644 after 13 years of being investigated by the Inquisition (including imprisonment and house arrest).

Until the end of the eighteen century the nature of the atmospheric air was not known. For instance, Antoine Lavoisier (1743-1794), the prominent eighteenth century chemist wrote: *"Are there different kinds of air? Are the different airs that nature offers us, or that we succeed in making, exceptional substances or modifications of the atmospheric air?* "The following discussion shows how the composition of the atmospheric air was discovered.

"FIXED AIR" - CARBON DIOXIDE. In 1753, Joseph Black studied properties of limestone (chalk or calcium carbonate, $CaCO_3$) and magnesia alba (magnesium carbonate, $MgCO_3$). Black discovered that when limestone or magnesia alba reacted with acid, they produced a strange kind of "air". The new "air" could extinguish a burning piece of paper or suffocate small animals. Black found that this air was particularly evident in human breath and bubbles of fermenting beer, and therefore it was always present in the atmosphere. Because this mysterious gas could be absorbed (or "fixed") by derivatives of lime ($Ca[OH]_2$), or magnesium, and also by a variety of animal, vegetable and mineral substances, he named it *fixed air*. Today the same gas is known as *carbon dioxide*.

Joseph Black (1728-1799)

Joseph Black was born on 16 April 1728 in Bordeaux, France into a Scottish Catholic family of twelve children. He studied medicine at the University of Glasgow, where he attended the lectures of William Cullen. Eventually, he became Cullen's assistant. Later, he taught chemistry at the University of Glasgow. His lectures were so popular that even students with no particular relish for the subject wanted to attend them. He was a pale, tall, thin, elegant, gentle man with dark eyes. He wore black speckled clothes, silk stockings, silver buckles, and carried either a slim green umbrella, or a genteel brown cane. He suffered from a chronic heart condition. Black never married. In 1766, he moved to Edinburgh to become a professor of chemistry at the University of Edinburgh. His main accomplishments were in quantitative chemistry and in discovering latent heat and specific heat. He died in Edinburgh on 6 December 1799.

Carbon dioxide is a colorless and odorless gas. When dissolved in water it produces a sour carbonic acid (H_2CO_3), presently used in beer or soda. Carbon dioxide is part of a vast planetary cycle, in which carbon circulates among three active reservoirs and undergoes several chemical changes. These reservoirs are the atmosphere, land, and the oceans. Carbon dioxide enters the atmosphere mainly from the decay of vegetation, but also from volcanic eruptions, the burning of fossil fuels, such as coal, oil, or natural gas, and from exhalations of animal life.

Chemical experiments. There are many unexpected advantages in performing simple chemical experiments. For example, in 1856, an 18 year old student, William Henry Perkin (1838-1907), performed some experiments at home, after his classroom teacher mentioned how valuable it would be to produce synthetic quinine. Perkin failed to make quinine but, by accident, came up with a purple dye "aniline". Consequently, he opened a factory and became a millionaire. Nevertheless, it is wise to be careful when performing chemical experiments. Many scientists, such as the German-Swiss chemist, Christian Friedrich Schönbein (1799 -1868, discoverer of ozone), learned this lesson the hard way. One day in 1845, Schönbein was in the kitchen experimenting with a mixture of nitric and sulfuric acids, even though such activities were strictly forbidden by his wife (his wife was absent at the time). He was just finishing an experiment when he heard his wife returning home. Schönbein got excited and, as a result, spilled a few drops of a new mixture he had just obtained on the table. In a hurry to clean it up, he grabbed the first thing at hand, which was his wife's cotton apron. He quickly used it to wipe the table and then hung it by the fire to dry, relieved that he had finished the job before his wife entered the kitchen. However, his wife did not have to enter the kitchen to know that her husband had broken the rules. A few seconds later the house was shaken by a huge explosion, caused by the reaction of Schönbein's new mixture with the cotton material and triggered by the heat of fire. Incidentally, this is also how Schönbein invented smokeless gun powder!

2.3. MAKING CARBON DIOXIDE (1)

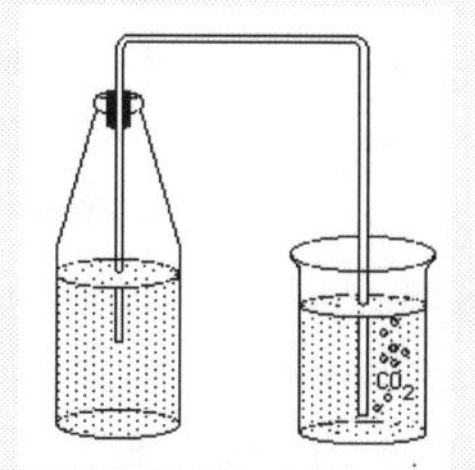

Materials: baking soda, 1/2 cup of vinegar, 1 liter bottle, stopper.

Procedure: Place 1/4 teaspoon of baking soda into a bottle. Slowly add vinegar. Seal the bottle and shake it. If the bottle is connected with a glass of water by a plastic tube, bubbles of CO_2 can be observed.

Explanation: Baking soda ($NaHCO_3$) is a chemical compound made up of the elements of sodium (Na), carbon (C), hydrogen (H), and oxygen (O). Vinegar is weak acid (CH_3COOH). When an acid is added to baking soda, it produces a chemical reaction and carbon dioxide (CO_2) is given off: $NaHCO_3 + CH_3COOH \rightarrow CH_3COONa + H_2O + CO_2$.

2.4. MAKING CARBON DIOXIDE (2)

Materials: chalk, 1/2 glass of vinegar, 1 liter bottle, candle, small glass.

Procedure: Place crushed chalk in a bottle. Slowly add vinegar and seal the bottle. Observe the bubbles of carbon dioxide which are forming in the bottle. Light a short candle and put it into a small glass. Unseal the test tube with carbon dioxide and "pour" the gas into the glass with the candle. Note that the flame will be put out.

Explanation: Chalk ($CaCO_3$) reacts with vinegar (CH_3COOH) and produces carbon dioxide: $CaCO_3 + 2CH_3COOH \rightarrow Ca(CH_3COO)_2 + CO_2 + H_2O$. Carbon dioxide is heavier than air and can be poured into the glass. It replaces air and as a result the flame of the candle is extinguished (since it cannot survive without oxygen).

The average person exhales about 0.9 kg of CO_2 daily. Since there are over 5 billion people on our planet, this means that all together they breathe out about 4,500,000,000 kg of carbon dioxide daily. The removal of carbon dioxide from the atmosphere takes place during the process of *photosynthesis*, when plants consume it to produce organic matter and oxygen. A large amount of CO_2 is stored in the oceans. This amount can vary since solubility of gases in water decreases with increasing temperature.

Two hundred years ago, the average annual concentration of CO_2 in the atmosphere was about 280 ppm (parts per million, 1 ppm = 0.0001%) in a volume of air. Since then, because of burning of coal and oil, the amount of carbon dioxide has increased exponentially. In 1994, the average annual concentration of CO_2 in the atmosphere was about 355 ppm. This increase has been of great concern to scientists because carbon dioxide (together with water vapor, methane, ozone, and nitrous oxide) absorbs the radiation emitted by the Earth. Therefore, carbon dioxide prevents part of the Earth's radiation from escaping into space and "keeps" the Earth warmer. This phenomenon is called the *atmospheric greenhouse effect* (see page 89).

The increase in the concentration of carbon dioxide in the atmosphere, due to industrial activities and forest depletion, can contribute to global warming. This in turn can further increase amounts of carbon dioxide in the air due to the reduction in CO_2 solubility in the oceans. Fortunately, some of the warming caused by the greenhouse effect is expected to be canceled at high altitudes by radiative cooling caused by scattering of solar energy to space by anthropogenic *aerosols* (tiny liquid and solid particles suspended in the air and produced in industrial activities).

The rising content of CO_2 in the atmosphere provides a strong impetus for plants and forest growth due to photosynthesis. As a result, expansion of plants could cause larger CO_2 removal from the atmosphere and might even prevent its concentration from rising further.

"INFLAMMABLE AIR" - HYDROGEN. In 1766, a British scholar Henry Cavendish (1731-1810) discovered hydrogen by treating metal with sulfuric or hydrochloric acid. He called the new gas *inflammable* *air* and thought of it as a pure phlogiston released from metal.

Antoine Lavoisier (1743-1794) showed that inflammable air could be produced by passing water, droplet by droplet, through a red-hot gun-barrel ($3Fe + 4H_2O \rightarrow Fe_3O_4 + 4H_2$). In an experiment performed in March 1784 in his Paris laboratory, Lavoisier produced 82 pints of inflammable air. Consequently, Lavoisier changed the name "inflammable air" to *hydrogen*, which simply meant "water former".

Hydrogen atoms make up about 90% of the universe. Hydrogen is not very abundant in our atmosphere. The reason is that hydrogen is about eleven times lighter than air, and therefore it is able to escape to space. In the atmosphere, most hydrogen is combined with oxygen in a form of water.

2.5. MAKING HYDROGEN (1)

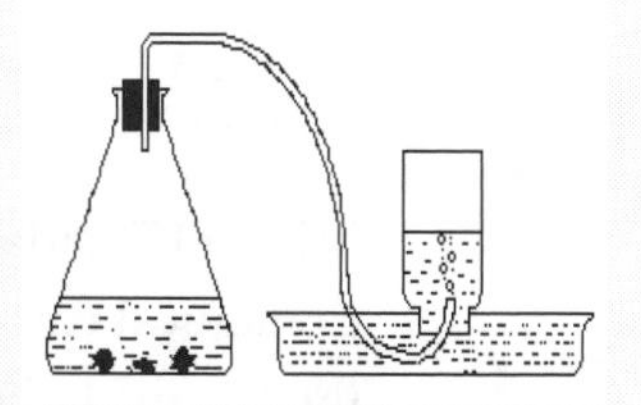

Materials: vinegar, iron wire, "Pyrex" glass test flask, stopper, plastic tube, jar, funnel, wood splint, hot plate.

Procedure: Pour vinegar through a funnel into the flask until it is partly filled. Cut the wire into small pieces and place them into the flask. Insert the stopper into the flask. Connect the flask with the inverted jar using a plastic tube. Heat the flask carefully until the liquid begins to boil. Bubbles of H_2 will be observed and collected in the jar. Light the wood splint, and insert it into the jar. The gas in the jar will pop. Be sure to keep the jar away from you because hydrogen mixed with oxygen is explosive. Since only a small amount of hydrogen is produced, this experiment is not dangerous.

Explanation: The iron is used to break up the vinegar (CH_3COOH) and release the hydrogen (H_2): $Fe + 2CH_3COOH \rightarrow Fe(CH3COOH)_2 + H_2$. The ability of hydrogen to explode with a pop (when a burning match or splint is brought close to it) is used to test for the presence of hydrogen.

2.6. MAKING HYDROGEN (2)

Materials: glass of water, two pieces of copper wire, flashlight battery (6V), two teaspoons of salt.

Procedure: Dissolve two spoons of salt in a glass of water. Form a little spiral at the end of each piece of the copper wire, by turning it around a pencil a few times. Attach the remaining ends of the wire pieces to the battery poles. Insert the spirals into the glass. Observe the bubbles of hydrogen appearing over the negative electrode. To collect it you might use a inverted test tube placed over the electrode

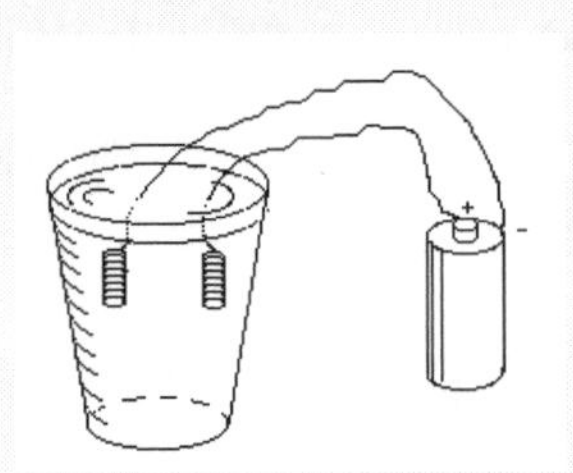

Explanation: The experiment is called *electrolysis*. It was first performed by Michael Faraday (1791-1867). Electrolysis is a method of splitting up a chemical compound by passing an electric current through it. Salt (NaCl) in the water splits into positively charged atoms (ions) of sodium (Na^+) and into negatively charged atoms of chlorine (Cl^-). The negatively charged electrode attracts the positive ions of sodium. After reaching the electrode, the ions of sodium give up their charges and immediately react with water: $2Na + 2H_2O \rightarrow 2NaOH + H_2$. Consequently, hydrogen comes from the negative electrode. The positively charged electrode attracts the negative ions of chlorine, which give up their charges and become chlorine gas. Chlorine reacts with water to form hydrogen chloride and oxygen, $Cl_2 + H_2O \rightarrow 2\ HCl + O$. As a result, chlorine and oxygen come from the positive electrode.

Atmospheric trivia. The atmosphere rotates with the Earth at a speed of about 1667.6 km/h (at the equator). Its depth is very small (about 1%) in comparison with the Earth's radius. About 50% of the atmospheric mass is located in the first 6 km of the atmosphere.

"NOXIOUS AIR" - NITROGEN. In 1770, Daniel Rutherford (1749-1819), a Scottish botanist (and student of Joseph Black at the University of Edinburgh), captured a sample of the atmospheric air in an inverted jar placed in a pan of water. In the jar, Rutherford burned a piece of phosphorus. Phosphorus totally absorbed the oxygen and produced fumes which were dissolved by the water in the pan. As a result, the level of water in the jar rose. In addition, caustic potash (potassium hydroxide, KOH) was used to absorb the fixed air (carbon dioxide) in the jar. The remaining gas, amounting to about three quarters of the volume of the original sample, instantly extinguished the flame. Rutherford labeled it *noxious air*. The gas discovered by Rutherford is now called *nitrogen*.

2.7. MAKING NITROGEN

Materials: test tube (or bottle), steel wool, jar.

Procedure: Place a ball of steel wool into a test tube. Invert the tube and submerge it in a jar with water. The oxidation of the steel wool will use up some of the oxygen in the container. Within a day or two, the water will rise to about one-fifth of the height of the tube. To remove the water vapor by condensation cool the tube by dropping ice cubes into the jar. The gas that remains in the test tube is 99% pure nitrogen.

Nitrogen is a colorless, odorless, tasteless, nontoxic gas, which is chemically not very reactive. Nitrogen is part of the planetary cycle, in which it circulates between the atmosphere and the Earth. When heated (solar radiation or lightning), molecules of nitrogen combine with oxygen and, as nitrates, enter the soil with precipitation. In soil, they become part of living organisms. When plants and animals die, the nitrogenous compounds are broken down into ammonia. Ammonia is converted by microorganisms into nitrates (salts or esters of nitric acid), nitrides (e.g., oxides) and pure nitrogen, which are returned into the atmosphere.

"DEPHLOGISTICATED AIR" - OXYGEN. A Swedish apothecary, Carl Wilhelm Scheele (1742-1786), and a British scholar, Joseph Priestley (1733-1804), independently discovered oxygen.

During Priestley 's experiment, performed on 1 August 1774, a piece of red mercuric oxide (HgO) was placed within a receiver inverted in a pneumatic trough and heated with a burning lens. An unknown gas was released during the process. Priestley first thought that the obtained gas was *noxious air* (nitrogen), but later changed his mind. He examined the gas and realized that a candle was brightly burning in it. Consequently, Priestley named the new gas *dephlogisticated air*, meaning lacking *phlogiston*. Now it is known that the gas he discovered was oxygen. Priestley found that *dephlogisticated air* was essential to the life of animals. For example, when he put a mouse into this gas, it could easily breathe. Priestley also discovered that in the sunlight, green plants gave off *dephlogisticated air*.

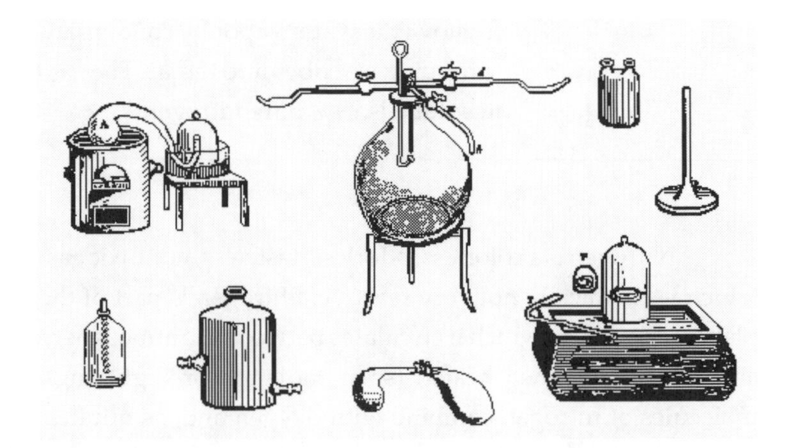

Early chemical equipment

Joseph Priestley (1733-1804)

Priestley was born on 13 March 1733 in Birsal Fieldhead, England. After his mother's death, he lived with an aunt, a childless widow, until he was 19 years old. He attended local parish schools and also was privately tutored. He learned Latin, Greek, Hebrew, Arabic, French, Italian, and German. In the 1770's, he lived near a brewery in Leeds. The brewery provided him with abundant carbon dioxide for experimentation. In 1772, he published a formula for "impregnating water with fixed air", which he hoped might be effective in curing scurvy. It proved to be ineffective, but nevertheless, the drink became popular. Priestley valued his discovery of the carbonated drink more highly than that of oxygen. In 1784, he was elected as a foreign associate of the Royal Academy of Sciences in Paris (and later in Boston, Philadelphia, Stockholm and St. Petersburg). Until 1789, he lived in London and then moved to his brother-in-law's house in Birmingham. He became known as a liberal critic of political discrimination against theological dissent in England, and a strong supporter of the French Revolution (1789-1799). On the second anniversary of the storming of the Bastille in Paris, a reactionary mob from Birmingham burned down his laboratory. As a result, Priestley moved to London. In 1794, he left England and moved to the United States, where he was warmly received and offered a position as a professor of chemistry at the University of Pennsylvania. He chose to spend the rest of his life in Northumberland, Pennsylvania. His youngest and favorite son died in 1795; his wife died in 1796. Ill and lonely during his last years, he died on 6 February 1804. The Smithsonian Institution in Washington, D.C. houses most of the remnants of his laboratory.

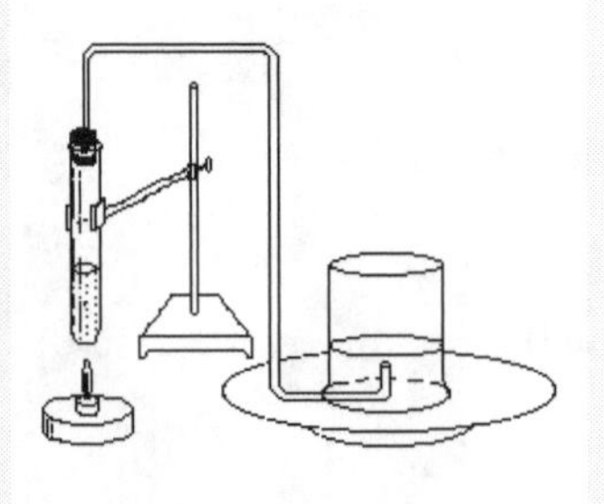

2.8. MAKING OXYGEN (1)

Materials: baking soda, hydrogen peroxide (usually available as 3% solution), test tube, test-tube holder, stopper, wood splint, jar, plate, plastic tube, source of heat.

Procedure: Pour hydrogen peroxide into a test tube until it is one-fourth full. Place one-fourth teaspoon of baking soda into the tube. Loosely insert the stopper in the mouth of the tube. Hold the tube with a test-tube holder so that the opening points away from you. Connect the test-tube with the jar using the plastic tube. Heat the tube until the liquid begins to boil. Light the splint and blow out the flame. Lower the glowing splint into the jar. The splint will burst into flame.

Explanation. Hydrogen peroxide (H_2O_2) is made up of the elements of hydrogen (H) and oxygen (O). It is a very unstable compound which easily breaks up into water (H_2O) and oxygen (O_2). Baking soda ($NaHCO_3$) helps to decompose the peroxide. When both are heated simultaneously, bubbles of oxygen are released. The glowing splint is used as a test for oxygen, since oxygen rich environment supports intense oxidation, with smoldering combustibles bursting into flame.

Chemical symbols. The letter system of chemical symbols, H for hydrogen, O for oxygen , etc., was introduced by the Swedish chemist, Jons Jacob Berzelius (1779-1848); prior to that each chemical element had an alchemic sign. For example, the alchemic sign of copper was: ♀

In October 1774, Joseph Priestley arrived to Paris and met with Lavoisier and other French chemists. During the visit, Priestley described his experiment with red mercuric oxide. Soon after Priestley returned to England, Lavoisier repeated the experiment in a reversible cycle and measured the quantities which were exchanged. First, he burned mercury, so that it combined with oxygen, and produced the oxide ($2\,Hg + O_2 \rightarrow 2\,HgO$). At the same time, Lavoisier measured the exact quantity of oxygen which was taken up from a closed vessel during the burning. Next, Lavoisier reversed the process by vigorously heating the mercuric oxide that had been made. Mercury was left behind and exactly the same amount of oxygen, which was absorbed before, was released into the vessel ($2\,HgO \rightarrow 2\,Hg + O_2$). During the experiment, the volume of oxygen which had been first absorbed and later released, was measured by the drop in the column of liquid. Based on his experiments, Lavoisier concluded in his *Traité élémentaire de chimie*, published in 1789:*"nothing is created either in the operations of the laboratory, or in those of nature, and one can affirm as an axiom that, in every operation, there is an equal quantity of matter before and after the operation, that the quality and quantity of the principles are the same, and that there are only alterations and modifications"*.

The best gun powder. In 1775, Lavoisier was appointed as one of four members of the Powder Commission. The Commission controlled the Powder Farm, a government corporation which had a monopoly on producing and selling gun powder. In 1788, Lavoisier applied his chemical genius towards improving the quality of gunpowder. The result was the production of the best gun powder in the world. The range of musket fire was increased from about 150 meters to about 250 meters. One might wonder how the superiority of French gunpowder, prepared by Lavoisier, influenced the victories of the Napoleonic Wars.

Antoine Laurent Lavoisier (1743-1794)

Antoine Lavoisier is regarded as the founder of modern chemistry. He developed quantitative methods of chemical analysis and pioneered the custom of naming chemicals after the elements they contained. Lavoisier was born on 26 August 1743 in Paris, during the reign of Louis XV. From 1754 to 1763, he attended the Collège des Quatre Nations. In 1768, Lavoisier bought a third interest share in Farmers-General, a private financial institution with a royal license to collect taxes on tobacco and salt, and custom duties on produce entering Paris. Later, these business dealings proved to have serious consequences. In 1771, the twenty-eight year old Lavoisier married Marie Anne Pierrete Paulze, the fourteen year old daughter of the farmer-general Jacques Paulze. The marriage was childless, but happy and harmonious. Mme. Lavoisier taught herself to be her husband's scientific assistant. She learned to speak English in order to help him with his translations. In addition, she studied drawing with the famous French painter Jacque-Louis David (who painted their portrait), and illustrated some of Lavoisier's works. She was also a hostess for the weekly gatherings of his scientific colleagues. In 1785, Lavoisier was appointed by the King to the directorship of the Academy of Sciences. He participated in the works of a commission which established the metric system of measures. In 1791, he was appointed to the board of the newly organized National Treasury. During the French Revolution, despite his respected contribution to the sciences and France, Lavoisier was violently attacked by militant journalists such as Jean-Paul Marat. On 24 December 1793, during the Reign of Terror, he was arrested together with his father-in-law and all other farmer-generals. For his association with a tax-collecting institution, he was tried on the morning of 8 May 1794, and was sentenced to be executed at the guillotine. During the trial, a report from the Bureau of Arts and Crafts was presented as evidence of Lavoisier's scientific contributions. The judge glanced at it and announced: "*The Republic has no need for scientists*". Antoine Lavoisier's life ended around 6 PM on 8 May 1794. He was but one of twenty eight prisoners guillotined that day. Lavoisier was fourth in line to climb the steps of the scaffold. The famous mathematician, Joseph Louis Lagrange, reacted to the news of Lavoisier's death by saying: "*It took them only a moment to cause his head to fall, and perhaps a hundred years will not suffice to produce its like!* "

Lavoisier showed that *dephlogisticated air* made up about a fifth of the air. He also invented a new name for it -- *oxygen*, a term derived from the Greek word for "acid producer". However, it turned out to be for the wrong reason. He mistakenly thought that the new gas was the basis of all acids (the nature of acids was discovered about 40 year later).

Lavoisier and his assistants perform experiments on respiration.

About 1772, a Paris jeweler named Maillard informed Lavoisier that a diamond would not burn in the absence of air. Lavoisier performed an experiment to confirm this notion. For several hours, he intensely heated three diamonds, furnished by Maillard, in a clay pipe filled with powdered charcoal. The diamonds did not burn. Based on this and other experiments performed during the following years, Lavoisier proved that oxygen played a vital role in the burning processes by combining chemically with the substance being burned. He also demonstrated that rust was not a metal *minus* "something", as the phlogiston doctrine asserted, but a metal *plus* "something", and this "something" was oxygen. As a result, the phlogiston theory was demolished.

The grave encounter. In 1780, Jean-Paul Marat (1743-1793), a Swiss medical doctor, who had a lucrative medical practice in Paris, published a treatise *Recherches physiques de la feu* ("Physical Researches on Fire"). Marat submitted his treatise to the French Academy of Sciences, hoping to be elected as one of its members. Marat's conclusion contradicted Lavoisier's theory of combustion. As a result, Lavoisier opposed it on the basis that no valid experimental evidence had been provided. Soon after, a Paris journal published a notice that the treatise had been formally accepted by the Academy. Lavoisier, as the President of the Academy, publicly denied it. Thereafter, Marat despised Lavoisier. They met again twelve years later. Marat became a radical Jacobin leader of the French Revolution, responsible for the September-1792 Massacre, in which well over 1200 people died. On 27 January 1791, Marat's newspaper *L'Ami du Peuple* ("The Friend of the People") launched the first attack on Lavoisier, which as we already know, proved to be fatal.

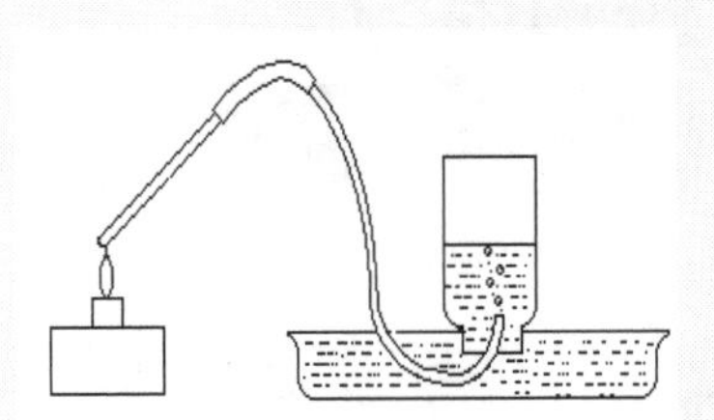

2.9. MAKING OXYGEN (2)

Materials: test tube, potassium permagnate (available in drug stores), jar, plate, wire.

Procedure: Place a teaspoon of potassium permagnate into a small test tube. Place the second test tube in a jar over water in a plate. Connect both tubes by a delivery tube. Thoroughly heat the tube with potassium permagnate. The produced oxygen gas will collect in the jar over the water in the plate. Hold the end of a piece of wire in the hot flame until it begins to glow. Quickly lower it into the jar containing oxygen and observe how the iron wire burns.

Explanation. Potassium permagnate ($KMnO_4$) is a chemical compound made up of the elements of potassium, manganese and oxygen. It will break up and release oxygen after being heated: $2KMnO_4 \rightarrow K_2MnO_4 + MnO_2 + O_2$.

2.10. OXYGEN IN THE AIR

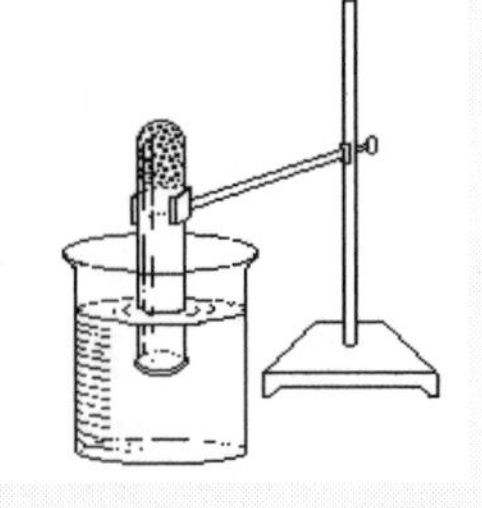

Materials: test tube (or bottle), steel wool, jar, test-tube, holder.

Procedure: Invert the test tube with steel wool and submerge it in a jar of water. Mark the water level in the test tube. After a few days you will see the red patterns of rust on the iron filings. The rusting (oxidation) of the iron filings will use up some of the oxygen in the jar. As a result, the water level will rise in the tube.

Explanation. The oxygen is removed by reacting with iron and forming iron oxide. The water column rises about one-fifth of the height of the tube because air is made up of about one-fifth oxygen. If there is not enough steel wool in the tube, the rise may be smaller.

Oxygen is a colorless, odorless, and tasteless gas. It is highly reactive and capable of combining with most elements. In the atmosphere, oxygen is present in the form of two-atom molecules O_2, three-atom molecules O_3, called *ozone* (from the Greek word "smell", because of its peculiar odor), and also one-atom molecules in the upper atmosphere. Ozone was discovered by Christian F. Schönbein (1799-1868) in the 1840's, but its existence in the air was not confirmed until 1913. Most of the ozone is found in the higher atmosphere, about 30 km above the Earth's surface. At this height, ozone is produced naturally and forms the so-called *ozone layer.* The ozone layer absorbs most of the ultraviolet radiation of the Sun. Ultraviolet radiation is harmful to life. Therefore, the presence of the ozone layer in the atmosphere protects life on Earth against radiation.

In 1974, two American scientists, Mario Molina and Sherwood Rowland, determined that certain man-made substances have been destroying the ozone layer. Among these substances were chlorofluorocarbons (CFCs) from spray cans and air-conditioning systems. As a consequence of the depletion of ozone, an increase in skin cancer in humans was expected, as well as various adverse effects on plants and animals. As a result of Molina's and Rowland's work, under the 1987 international agreement, called the Montreal Protocol, and a new agreement in June 1990, over forty industrialized countries pledged to completely eliminate the use of ozone-depleting chemicals by the year 2000. For their research, Molina and Rowland obtained the 1995 Nobel Prize in Chemistry.

OTHER ATMOSPHERIC GASES. In 1892, a professor at the Royal Institution in London, Lord Rayleigh (John William Strutt, 1842-1919), found that pure nitrogen obtained from ammonia weighed about 0.5% less than the nitrogen obtained from the air. This fact indicated to him the presence of another unknown gas in the air. In 1894, Rayleigh and a fellow chemist, William Ramsay, succeeded in isolating a small amount of this gas, which they named *argon* (which means "inactive"). They also proved that the new gas made up less than 1 % of the atmospheric air. For his research, which included the measurements of properties of the atmospheric gases, Rayleigh was awarded the 1904 Nobel Prize in physics.

Spectroscopy. The method used by Rayleigh and Ramsay to identify argon was called *spectroscopy*. The prototype of a *spectroscope* was built by the 19th Century German optician, Joseph Fraunhofer. The technique of spectroscopy was initiated in 1859, when two physicists, Robert Bunsen and Gustav Kirchhoff, discovered that light passing through different gases produced different bands of colored light called *spectra*.

During an 1868 eclipse of the Sun, the French astronomer, Pierre Jules Cesar Janssen (1824-1907), observed a bright yellow line in the Sun's spectrum. The line was attributed to an unknown element present in the Sun, and named *helium*, after the Greek word *helios* meaning "the Sun". During the following year, William Ramsay analyzed the Earth's minerals and identified the same element. Later, Ramsay in collaboration with the British chemist Morris W. Travers (1872-1961), used a spectroscope to isolate three other gases found in the atmosphere: *krypton, neon, and xenon.*

GAS MOLECULES. In 1808, the French chemist Joseph Louis Gay-Lussac (1778-1850) discovered an interesting regularity in the combination of gases. For instance, by combining hydrogen and oxygen, while measuring the volumes involved, he found that twice as much hydrogen reacted than oxygen (i.e. 2 volumes of hydrogen + 1 volume of oxygen resulted in 2 volumes of steam). After a series of experiments with other gases, he formulated the law: "*Under identical conditions of temperature and pressure, the ratios of volumes of gases participating in chemical reactions are small whole numbers*".

This finding puzzled the Italian professor of physics, Amadeo Avogadro (1776-1856). In order to explain it, in 1811, Avogadro concluded: "*the number of molecules in any gas is always the same for equal volumes, or always proportional to the volumes*". Avogadro also realized that if two volumes of hydrogen united with one volume of oxygen, it was only possible if each molecule of oxygen was divided between two molecules of hydrogen. This posed a difficulty because it was believed that molecules could not be split. To overcome this inconsistency, Avogadro postulated compound molecules. He formulated this conclusion as his second hypothesis: "*The particles of elementary gases do not need to be atoms, they may be molecules containing two or more identical atoms*".

Assuming that molecules of hydrogen and oxygen have two atoms, Avogadro argued that two molecules of hydrogen reacts with one molecule of oxygen: $2H_2 + O_2 \rightarrow 2H_2O$.

Humid air is lighter. When a number of water vapor molecules is added to a fixed volume, the same number of air molecules must leave this volume to keep the total number of molecules (Avogadro's law) as well as temperature and pressure constant. Assume that 10 moles (1 mole is a fixed number of molecules defined on the next page) of water vapor replace 10 moles of air, i.e., 8 moles of nitrogen and 2 moles of oxygen. Molecular weights of water vapor, nitrogen, and oxygen are 18 g, 28 g, and 32 g, respectively. Therefore, the atomic weight budget is: (10 x 18 g) - (8 x 28 g + 2 x 32) = -108 g (lost). This indicates that humid air is lighter than dry air of the same temperature and pressure.

Many years after his death, Avogadro was honored by having a physical constant named after him. The constant provided the number of gas molecules (numerically, 6.02×10^{23}) contained in one *mole*. One *mole* occupies 22.4×10^3 cm^3 (at standard conditions of temperature and pressure).

ABUNDANCE OF ATMOSPHERIC GASES. The table below shows the volume percentages of the various gases (permanent and variable) present in the Earth's standard atmosphere (below 80 km):

Permanent Gas	Symbol	% by volume	Variable Gas	Symbol	% by volume
Nitrogen	N_2	78.08	Water vapor	H_2O	0-4
Oxygen	O_2	20.95	Carbon dioxide	CO_2	0.0351
Argon	Ar	0.93	Ozone	O_3	0.000004
Neon	Ne	0.0018	Carbon monoxide	CO	0.00002
Helium	He	0.0005	Sulfur dioxide	SO_2	0.000001
Methane	CH_4	0.0001	Nitrogen dioxide	NO_2	0.000001
Hydrogen	H_2	0.00005			
Xenon	Xe	0.000009			

Nitrogen and oxygen are the most abundant gases in the atmosphere. Their total amount is about 99.03% of dry air. Concentration of water vapor, one of the variable atmospheric gases, changes in time and space as a part of the natural hydrologic cycle. Concentration of CO_2 shows cyclic oscillations associated with the annual vegetation cycle. Amount of SO_2 in the atmosphere may vary due to volcanic eruptions into the upper atmosphere. Some of the variable atmospheric gases change their amounts in the atmosphere as a result of *anthropogenic* (man-made) industrial activities. For example, *ozone*, *nitrogen dioxide*, and *carbon monoxide* are emitted into the atmosphere by motor vehicles, as a result of the high-temperature combustion of fuel. *Carbon dioxide* and *sulfur dioxide* are produced by burning wood and coal. Sulfur dioxide (SO_2) readily oxidizes to *sulfur trioxide* (SO_3). Sulfur trioxide in moist air reacts with water and produces *sulfuric acid* (H_2SO_4). Sulfuric acid can be transported within clouds for hundreds of kilometers. When it is removed from the clouds, it results in *acid rain*.

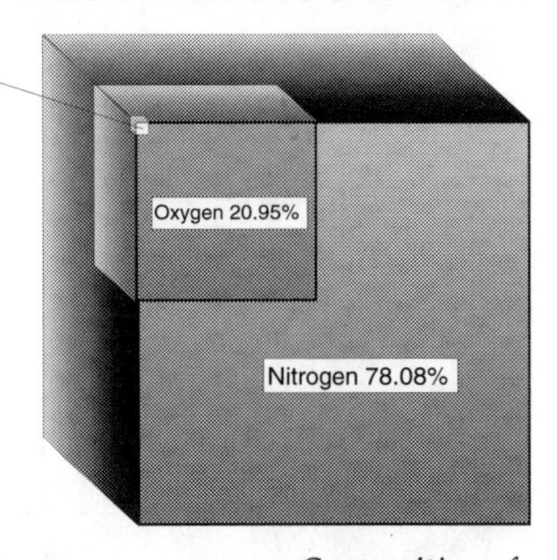

Composition of the atmosphere

Among all of the gases present in the atmosphere, oxygen, naturally produced ozone, water vapor, nitrogen, and naturally produced carbon dioxide have the most important function of maintaining life on Earth. Without oxygen in the atmosphere, we would die of suffocation. Without ozone, we would die of sunburn. Without water (vapor), we would die of thirst. Without carbon dioxide and nitrogen, we would die of starvation.

STRUCTURE OF THE ATMOSPHERE. The vertical structure of the atmosphere was first investigated through the use of balloons. The first manned flights were performed in November of 1783 by the Montgolfier brothers, French paper manufacturers at Annonay who used a hot air balloon made of paper.

Jacques Charles (1746-1823), the French professor of physics, used hydrogen to fill a large (9 m in diameter) rubberized silk balloon named "Globe". On 1 December 1783, Charles, accompanied by Nicholas Roberts, made the first manned flight with this type of balloon. In order to produce the required amount of hydrogen, Charles used 540 kg of iron filings and 270 kg of diluted sulfuric acid. It took four days to inflate the balloon. The launch took place at the garden of the Tuileries. After the 25 minute flight, Charles and Roberts landed in a meadow, 27 miles northwest of Paris. Later, only Charles ascended again and reached an altitude of more than 1000 m, where he measured a temperature of -5°C, 13 degrees lower than the surface temperature.

On 27 August 1804, two French scientists, Jean Biot and Joseph Louis Gay-Lussac, ascended in a hydrogen filled balloon

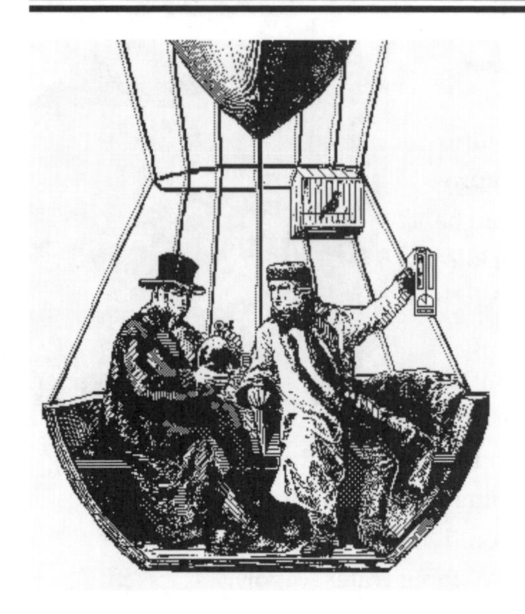

Biot and Guy-Lussac during their balloon flight

for the first scientific reconnaissance flight. Biot and Gay-Lussac flew to the level of 7 km and made observations about the nature of the atmospheric air. They carefully noted the effects of the altitude on the animals which accompanied them on the flight. They also took a sample of air for further analysis. The analysis led them to conclude that the composition of the atmosphere does not change with height. The same amounts of nitrogen and oxygen were found aloft as at the surface. Consequently, for the first time air was recognized to be a remarkable mixture in which proportions of the constituents remain constant at any given altitude (up to about 80 km). Three weeks later, Gay-Lussac repeated this ascent alone, carrying flasks for air samples. He also made observations of how temperature changed with height.

Between 1899 and 1902, the French scientist, Leon Philippe Teisserenc de Bort (1855-1913), performed a three year study using small unmanned balloons filled with hydrogen. He tracked these balloons with two *theodolites*, instruments used by surveyors to measure horizontal and vertical angles. The research led de Bort to conclude that atmospheric properties, especially temperature, changed with height. De Bort named the lower portion of the atmosphere, where temperature decreases with height, the *troposphere*, from the Greek words *tropos* meaning "turn" and *sphaira*, meaning "ball". He also called the upper portion of the atmosphere, where he expected a stratum of light gases to exist, the *stratosphere*.

With time, unmanned balloon soundings became the routine method for securing measurements in the upper atmosphere. The standard measuring system consists of a helium or hydrogen balloon, and an instrument package, called *a radio-*

sonde. The first expendable radiosondes were designed and tested in 1939 by the Russian meteorologist, Pavel A. Molchanov.

A radiosonde can ascend to 30 km measuring pressure, temperature, humidity, and transmitting these data via radio to the ground station. Radiosondes are released twice daily by 950 stations throughout the world as a part of a routine international program of the World Meteorological Organization (WMO).

Record balloon flights. The highest scientific balloon flight was performed in 1960 by Captain Joseph W. Kittinger, Jr. of the United States Air Force. In an open gondola balloon, Kittinger rose to a level of 34.3 km above the New Mexico desert. After reaching this altitude, the pressure-suited aeronaut jumped out. At 6 km above the ground, when Kittinger's falling speed was 380 km/h, his main parachute opened automatically. Nine minutes later, he landed safely on sand.

Two decades later, a young American attempted to fly even higher on a "craft" consisting of several hydrogen-filled balloons. He reached an elevation of about 10 km, where the air temperature registered below -30°C. Almost frozen, the young man could barely manage to shoot some of his balloons so that he could return to Earth. As if this wasn't adventure enough, the police were awaiting his return so that they could book him for violating US air space regulations.

Research performed with radiosondes and meteorological rockets revealed that the Earth's atmosphere can be divided into a series of layers, based on its thermal structure. The *troposphere* is the lowest, the thinnest (about 10 km), but the most dense layer of the atmosphere. Above, in the *stratosphere*, the temperature of the air increases because of the absorption of solar radiation by the ozone layer. The transition layer, sandwiched between the troposphere and the stratosphere, is called the *tropopause*. Meteorologists have learned that the stratosphere extends to a height of 50 km. The transition layer, above the stratosphere, is called the *stratopause*. Above this layer there is another one, called the *mesosphere* (from the Greek word:

"middle"), which extends to a level of 80 km. The temperature decreases through the mesosphere, where most of the meteorites burn up. The transition layer above the mesosphere is called the *mesopause*. Above it, there is a layer called the *thermosphere* (Greek for "heat"). In the thermosphere, solar energy raises the air temperature (i.e., the velocities of air molecules). The density of air at this level is practically zero. Therefore, the outdoor thermometer would not be functional at such high altitudes.

Lost balloonists. Once, during cloudy conditions, two hot-air balloonists lost their direction while in a flight over Colorado. As they were slowly drifting over the grounds of a well-known meteorological research institution, one of the balloonists suddenly spotted a man walking below and yelled to him:

- *Sir, could you tell us where we actually are?*

After a few moments the man on the ground answered:

- *You are in a hot-air balloon!*

The aeronauts looked at each other discontentedly, and one of them mumbled:

- *The fellow down there obviously must be a scientist, his reply is so perfectly precise and so utterly useless!*

ORIGIN OF THE ATMOSPHERE. Scientists believe that billions of years ago the Earth's atmosphere was entirely different. The primitive atmosphere included hydrogen, helium, and other simple gases. This atmosphere was lost to space because the constituents were too light and mobile. During the first stages of the Earth's existence, countless collisions with meteorites generated tremendous heat on its surface. This heat melted a great deal of the Earth's rocks. During these processes many gases were emitted from the molten Earth. After a certain time, the Earth began to cool and solidify. Molten rock occasionally broke through the Earth's surface in the form of volcanoes and continued to emit gases. The accumulation of these gases above the surface formed the secondary primitive atmosphere.

Earth's prehistory. The oldest rocks found on the Earth are thought to be about 4 billion years old. Some meteorites are probably as old as 4.5 billion years. Therefore, the Earth and the solar system are believed to be formed about 4.5 billion years ago. The oldest-known fossil plant on the Earth, the single-celled blue-green algae, comes from rocks found in South Africa that are about 2 billion years old. Blue-green algae still exist throughout some parts of the world and form colonies that form limestone.

Contemporary volcanoes probably produce the same gases, and in similar proportions, as they did millions of years ago. It can be imagined that volcanic activities in ancient times generated carbon dioxide, carbon monoxide, water vapor, nitrogen, and sulfur dioxide. The early atmosphere contained virtually no oxygen. This finding follows from the fact that some of the oldest rocks which were exposed to the early atmosphere are deficient in oxygen. If oxygen had been present in the atmosphere, it would have combined with the iron in ancient rock deposits. Most of the oxygen which first appeared in the atmosphere was produced through *photosynthesis* by one-cell creatures called *blue-green algae,* which lived in the ocean three billion years ago. As oxygen accumulated in the atmosphere, certain amounts of ozone also formed and served as a protective layer from the Sun's ultraviolet radiation. Oceanic water acted as an additional shield against deadly cosmic radiation. As a result, life on Earth began below the ocean's surface, where there was ample protection. Once the ozone layer becomes sufficiently extensive, then plants and animals could safely leave the ocean and habituate the continents.

PAST CLIMATES. Climate is a description of the dynamic state of the earth-atmosphere system over a long period of time (years, decades, centennials, millennia, etc.), and over large areas (continents, the globe). Climate is determined by such factors as the Sun-Earth interactions, changes in the Earth's orbit, arrangements of continents and oceans, volcanic eruptions, anthropogenic (man-made) effects, etc. Scientists analyze climatic changes of the past and also study possible future projections.

Past climates are investigated by examination of biological (fossil plants and animals), chemical (bubbles of air in ice caps), and geological (ocean floor sediments) clues. Based on such hints, some scientists hypothesize that about 65 million years ago an asteroid, perhaps the size of Halley's comet, smashed to the Earth at more than 150,000 km/h. The energy of the impact locally increased temperature by 20,000 degrees. As a result, 200,000 km^3 of soil and rock instantly vaporized, forming the Chicxulub crater, 300 km across, in the Gulf of Mexico. The resulting dust veil might have reduced the amount of solar radiation and caused significant cooling on the Earth. The collision probably also sent 100-m tidal waves across the ocean and triggered earthquakes hundreds of kilometers away. There is also a belief that the resulting climate changes caused a massive extinction of large animals such as dinosaurs.

Climatologists also ascribe the Earth's climatic changes in the past to three separate cyclic variations in its orbit. According to the theory proposed in the 1930's by the Yugoslavian astronomer, Milutin Milankovich (1879-1958), the first cycle (about 100,000 years long) is due to the changes in the shape of the Earth's orbit. During this cycle, the orbit becomes more and less circular. During the second cycle (about 41,000 years long), the tilt of the Earth's axis changes from 22.5° to 24.5° (presently, it is 23.5°). The third cycle (about 22,000 years long) is caused by spinning-top-like wobbles of the Earth's axis in space. As a result of all these changes, the Earth's orientation towards the Sun varies, which also affects the climate.

The Earth' climate in the past may have been changed by a shifting of the continents. According to the theory of plate tectonics, the Earth's outer shell is composed of huge movable plates. Continents are embedded in the plates and move with them. Before breaking apart, all the continents were joined together in a single huge island called "Pangaea" and were surrounded by an ocean called "Panthalaasa", as proposed in 1912 by the German geologist Alfred L. Wegener (1880-1930).

It is by logic that we prove, but by intuition that we discover.
(J. H. Poincaré)

CHAPTER THREE

PRESSURE

Vacuum, Invention of the Barometer, Atmospheric Pressure, Power of Atmospheric Pressure, Pressure Laws, Pressure Units, Altitude Effects, Types of Barometers.

VACUUM. Pressure and vacuum are presently considered elementary phenomena, but to ancient scholars these were objects of profound philosophical discussions. Many Greek philosophers, including Thales and Plato, rejected the idea of a vacuum. Aristotle, in the fourth century B.C., also argued that *nature abhors a vacuum*. As a result, common suction pumps were thought to work because nature pulled up the water in order to prevent a vacuum from forming. This belief was held for many centuries.

Galileo Galilei (1564-1642) demonstrated that a vacuum can exist. He constructed a device which pulled down a

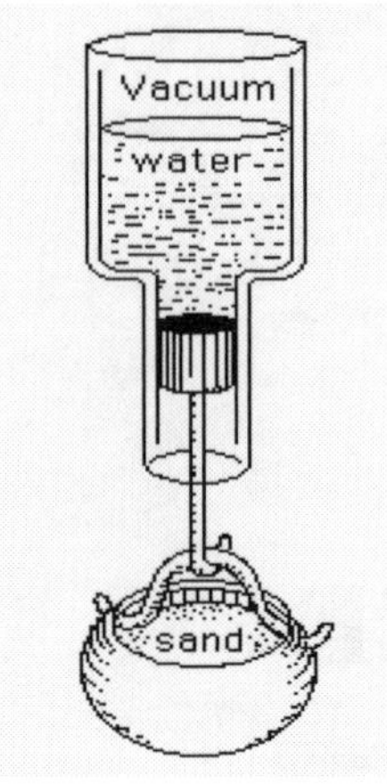

Galileo's experiment

wooden stopper in a thick-walled glass cylinder filled with water. As a result, he observed a space which separated the glass cylinder wall and the water level, and assumed that it was a vacuum.

In his treatise, *Discorsi*, published in 1638, Galileo attempted to explain that water could not be raised more than about 10 m by a lift pump, a fact long known to well diggers and miners. In trying to interpret this phenomenon, Galileo incorrectly compared water to a copper wire which would break of its own weight, if it were sufficiently long and thin.

Berti's experiment

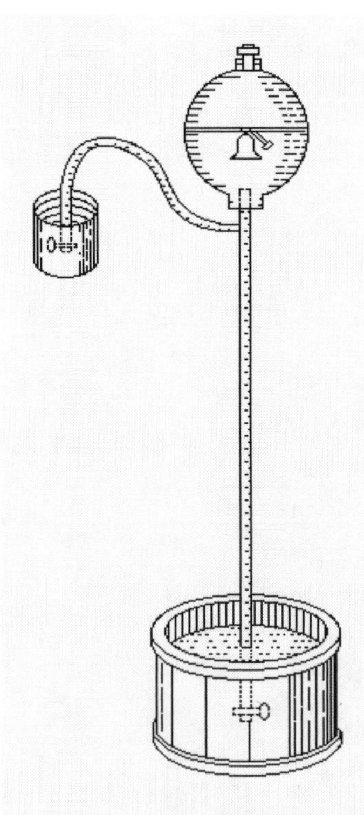

Galileo's *Discorsi*, which arrived in Rome in December of 1638, motivated a scholar named Rafaello Magiotti to suggest an experiment on the vacuum. The experiment was carried out by Gasparo Berti (1600-1643) in June, 1641. Berti used an apparatus which consisted of a lead tube about 12 m high, bent downward at the top and terminated at either end by a valve. Each valve was submerged in a container filled with water. During the experiment, the tube was filled with water, with the lower valve closed and the upper one opened. Next, the upper valve was closed, and the lower one opened. As a result, the level of water in the pipe dropped and created a vacuum. After the upper valve was opened, the air rushed in with a loud noise, and filled the space previously abandoned by the water. Later, to further test for a vacuum, a bell and a hammer were placed in the upper container and operated by an external magnet (sound does not travel through a vacuum).

INVENTION OF THE BAROMETER. Rafaello Magiotti described Berti's experiment in a letter to Evangelista Torricelli (1608-1647), the court philosopher in Florence. The letter suggested that if sea water were used, the level in the tube could be different. This perhaps inspired Torricelli to replace the water with mercury, a liquid metal 13.6 times more dense than water. Mercury was called *quicksilver* at that time, and was mined in the Italian province of Tuscany.

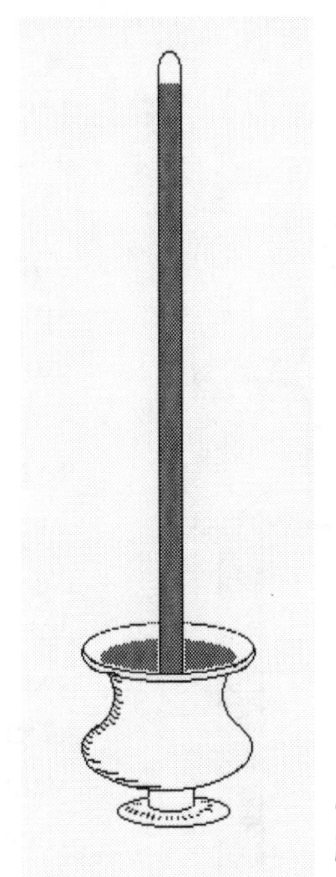

The first barometer

In 1644, a famous experiment was conducted by Vincenzo Viviani (1622-1703), Torricelli's assistant, at the suggestion of his master. A glass tube, about 1 meter high was filled with mercury, plugged and turned upside down. The plugged end was submerged in a basin of mercury, and finally unplugged. The volume of mercury dropped slightly but remained in the tube, about 76 cm above the level of mercury in the basin. This is how the barometer was invented.

To show that the space in the tube above the mercury was completely empty, Torricelli performed an additional experiment. A tube containing mercury was placed in a basin which was partly filled with mercury, while the water added to it remained on top. Little by little, Torricelli raised the tube to the water level. At the same time the water surged into the tube and filled it to the top. This was proof that the force holding the mercury in the tube must have come from outside. Torricelli concluded that "*noi viviano sommersi nel fondo d'un pelago d'aria elementare*", "we live submerged at the bottom of an ocean of elementary air". The cumulative weight of the layers of air in "*the 50 mile atmosphere*" (as Torricelli thought), exerted pressure and supported a column of mercury 76 cm high.

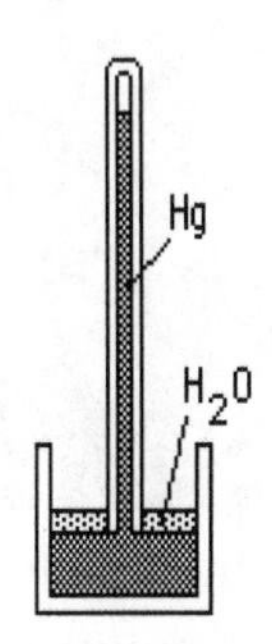

Torricelli's experiment

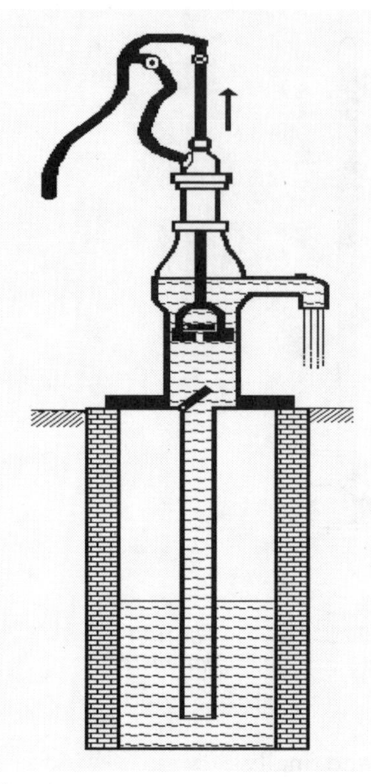

Suction pump

Based on the same principle, Galileo's problem of the suction pump was finally solved. The pumping action created a pressure deficit. This deficit could not exceed the value of the atmospheric pressure. The value of the atmospheric pressure was equivalent to the pressure exerted by a column of about 10 m of water. Therefore, water could be pumped only to a level of 10 m above its surface, but not any higher.

Torricelli did not publicly announce his discovery. Nevertheless, he described the experiment in a letter to his friend, Michelangelo Ricci, on 11 June 1644. Later that same year, Ricci wrote a letter to the French priest Pierre Martin Mersenne in Paris, revealing details of Torricelli's experiment. Mersenne, with his extensive correspondence, spread the news throughout Europe.

Evangelista Torricelli (1608-1647)

Evangelista Torricelli was born on 15 October 1608, near Faenza, Italy. Orphaned in his early childhood, he was reared by his uncle, the scholarly monk Jacopo. Torricelli first studied under the Jesuits in Faenza. In 1627, he went to Rome, where he studied under Benedetto Castelli, a friend of Galileo Galilei. Soon he became recognized as a talented mathematician. In 1641, he was nominated to be Galileo's assistant and moved to Florence. Galileo died three months later. After Galileo's death, the Grand Duke Ferdinand II appointed Torricelli to succeed Galileo as the court philosopher and mathematician in Florence. In 1644, Torricelli published his main work *Opera Geometrica*. He died on 25 October 1647, after a short illness, probably typhoid fever.

3.1 MAKING A BAROMETER

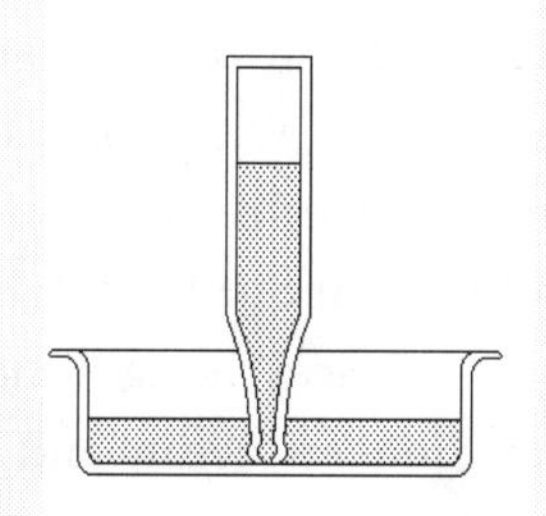

Materials: bottle, water, oil, saucer.

Procedure: Partly fill a bottle with water. Place the bottle vertically in an upside down position in a saucer filled with water. Put some oil on the water's surface in the saucer to prevent evaporation. Variations in atmospheric pressure will change the level of water in the bottle. These changes can be recorded on a strip of paper affixed to the outside of the bottle.

Comment: Notice that the readings of this instrument are also dependent upon the temperature of the air inside the bottle. In a room with a fairly uniform temperature of a room the instrument will work as a crude barometer.

ATMOSPHERIC PRESSURE. Many contemporaries became fascinated by the new instrument. For instance, in 1645, Torricelli's experiment was repeated privately in Rome for Ferdinand II's brother, Cardinal Giovanni Carlo. Tubes of widely different shapes and sizes, pointing in various directions, were used. Because the volume in each tube above the mercury was different, a conclusion was reached that changes in the level of the mercury in the tubes were caused only by atmospheric forces.

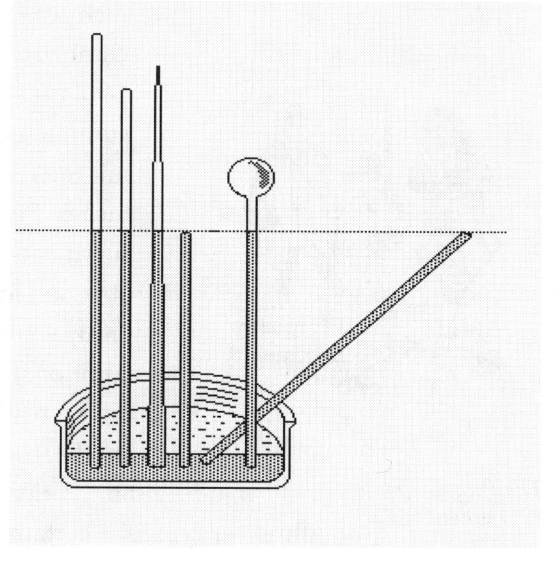

Pressure experiment

Blaise Pascal of France (1623-1662) repeated Torricelli's experiment with the assistance of his father, Etienne. He experimented with tubes of different shapes and lengths. Because he used red wine, which was lighter than water, the tubes had to be about 12 meters long.

Pascal thought that if Torricelli were correct, and if the air were as an "ocean" and had weight, then just as the cumulative weight of water increases with depth, so should the pressure of air decrease with height. Pascal realized that the easiest way to verify his supposition was to carry a barometer to the top of a mountain. However, because he was chronically ill, and lived in Paris, where there were no appreciable mountains, he was not able to personally perform the experiment. Consequently, he asked his brother-in-law, Florin Périer (1605-1672), a physicist, who lived in the mountainous region of Auvergene in southern France, to carry out his experiment.

The experiment was performed on 19 September 1648. Périer invited a few distinguished clerics and laymen from his town of Clermont-Ferrand to accompany him. First, they set up a barometer at the bottom of the mountain and measured the level of mercury in the tube at exactly 28 inches (the Parisian inch was 2.7 cm). Then, Périer and his party climbed the mountain (Puy de Dome, elevation 1467 m) and set up another barometer at the summit. In the thinner atmosphere of the summit the reading was lower and equal to 24 $^{2}/_{3}$ inches. Périer performed five additional measurements at various spots on the summit and obtained identical readings. As Périer and his group were descending the mountain, they took additional measurements and found that barometric readings were also increasing. Taking another reading after descending the mountain, Périer again obtained the value of 28 inches, thus verifying the certainty of his measurements. The results confirmed Pascal's belief that air was an elastic substance

The Puy de Dome experiment

whose pressure decreased with altitude.

Blaise Pascal (1623-1662)

Blaise Pascal was born on 19 June 1623 in Clermont-Ferrand, France. His mother died when he was three years old. He and his brothers and sisters were brought up by his father, Etienne. In 1629, the Pascal family moved to Paris. Blaise was soon recognized as a mathematical prodigy. In 1642, he invented the first mechanical adding machine. It could add and subtract, but was too expensive to be practical. His analysis of the cycloid inspired others to formulate the calculus. He also pioneered fluid mechanics. Pascal was once contacted by the Chevalier de Mèrè, a professional gambler. De Mèrè lost a sizable amount of money in a dice game and was interested in finding of a successful betting method. It is not known how de Mèrè fared in his profession, but Pascal's work on this problem did lead to the foundation of the theory of probability. In 1654, Pascal was almost killed by his own runaway horses. As a result of this incident, Pascal underwent a spiritual metamorphosis and turned to God, after having a mystic revelation. Pascal entered the Jansenist community, an extremely puritan sect within the Catholic church, condemned later by the Pope. His philosophical and religious ideas were expressed in his *Provincial Letters.* In the convent of Port-Royal des Champs, fifteen miles southwest of Paris, he lived a rigorously ascetic life . Pascal died on 19 August 1662. When his body was laid out for burial, it was discovered that he wore a spiked wire belt next to his skin.

Further research performed by Pascal led to the invention of a *hydraulic press* and the formulation of a hydrostatic law, known as Pascal's principle. According to this law, external pressure in liquids is redistributed equally in all directions. Pascal's results were published after his death in 1662, in a treatise *On Equilibrium of Liquids.*

A debate. At the end of 1647, barometric experiments had allegedly been conducted in Warsaw by V. Magni, who implicitly claimed priority. This fact created some debate among scholars. A few centuries later, in 1906, historian F. Mathieu accused Pascal of fabricating the letter from Périer of 15 November 1647, in order to take the credit for performing barometric experiments. After investigating the subject, another historian, J. Mesnar, concluded that Périer probably did send the contested letter to Pascal, who may have slightly changed the text for publication.

3.2. VERTICAL PRESSURE CHANGES

Materials: hammer, nail, can.

Procedure: Using the hammer and nail, make three holes of the same size in the side of a tall tin can. Punch one hole near the top, another hole in the middle, and the third one near the bottom of the can. Pour water from a faucet into the can. Keep the can full while the water is running out through the holes. Note that the greater the depth of the water, the greater the pressure at each hole, and the farther the water will spout out of the holes.

Mass of the atmosphere. In 1651, Blaise Pascal figured out that the total mass of the atmosphere is 8.28×10^{18} pounds. He did not reveal how he arrived at this calculation, admitting only that it was so elementary that *"a child who knew how to add and subtract could do it"*. Using a definition of pressure given in this Chapter and force (Chapter 7), it might calculated that the total mass of the atmosphere is 5.25×10^{18} kg, or 11.6×10^{18} pounds, quite close to Pascal's estimate.

3.3. PASCAL'S PRINCIPLE

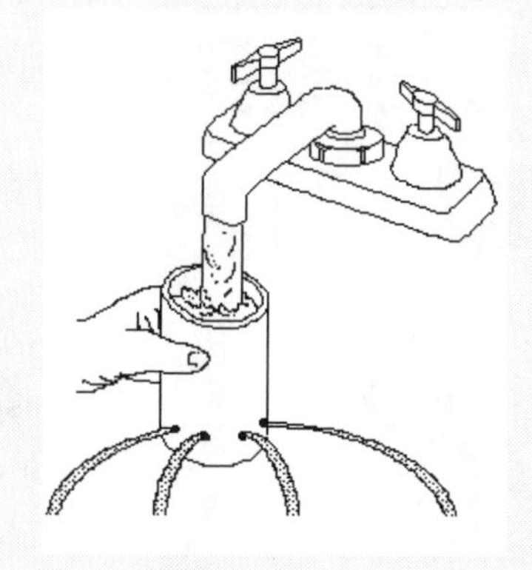

Materials: tin can, nail, hammer.

Procedure: Using a hammer and a nail, punch holes around the base of a tin can. Fill the can with water over a sink. Note how the water spouts out the same distance from all of the holes.

Explanation: According to Pascal's principle, pressure at a given point in a liquid is the same in all directions.

Atmospheric pressure and our bodies. According to a proverb, "when your joints all start to ache, rainy weather is at stake". Sensitivity to such changes varies from person to person. Those with scars or ailments are more sensitive and "feel it in their bones". Moreover, when pressure rises, the body absorbs air into its fluids to balance the compression. When pressure falls, the body has to remove air from its fluids to prevent swelling. This process is perceived by most people as not very pleasant. In addition, at low pressure, some of the sacks in joints expand and impinge upon nerves that cause the sensation of pain.

POWER OF ATMOSPHERIC PRESSURE. Otto von Guericke (1602-1686) was the first to demonstrate that the atmosphere exerts a rather tremendous pressure. In 1650, in order to prove the existence of a vacuum, von Guericke filled a sealed barrel with water, and then tried to pump it out with a suction pump he had constructed. The result was unsatisfactory because the wooden barrel was not strong enough. Then he replaced the wooden barrel with a copper sphere. It collapsed

Von Guericke's experiment

when he started pumping out the air. Von Guericke finally succeeded in maintaining a vacuum, by using a thicker sphere.

In another experiment, von Guericke intended to measure atmospheric pressure. He made two bronze hemispheres. One part had a valve on it which could be opened or closed at will. He pumped the air through the valve out of the two joined hemispheres. He attached the upper one to a wooden construction, and tried to pull the hemispheres apart by hanging various loads from a hook connected to the lower one.

In 1657, von Guericke performed a similar demonstration before the German Emperor, Ferdinand III, and the assembled Reichstag in Magdeburg. It took several horses to pull the hemispheres apart. This dramatically demonstrated the power of atmospheric pressure.

Magdeburg experiment

3.4. VACUUM (1)

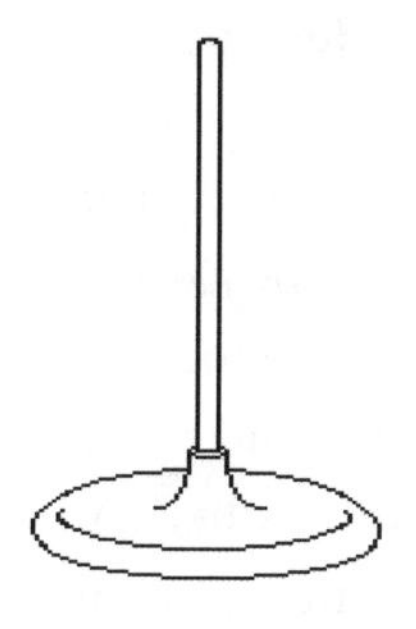

Materials: plunger, chair.

Procedure: Moisten the rubber cup of the plunger and press down firmly on a flat part of the seat of a chair until most of the air in the cup is expelled. You will be able to lift the chair with the plunger.

Comment: By forcing the air out of the cup, a partial vacuum was formed inside. The air pressure on the outside is greater than the air in the cup and holds the cup firmly to the chair.

3.5 VACUUM (2)

Materials: two plungers.

Procedure: Moisten the surfaces of both plunger cups. Hold one handle against the floor or the wall while you push one cup against the other to force the air out of both cups. Now try to pull the plungers apart.

Comment: This experiment is actually analogous to von Guericke's demonstration with the hemispheres. There is a partial vacuum inside the cups and the air pressure on the outside holds the cups firmly together.

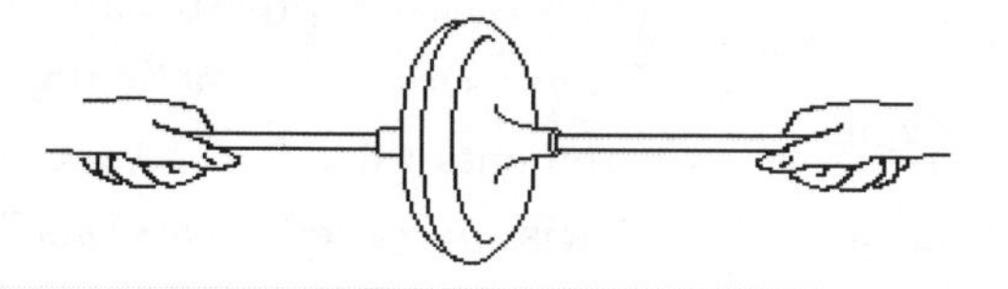

PRESSURE LAWS. The first laws governing atmospheric pressure were developed in England by Robert Boyle (1627-1691). Boyle was also credited for introducing the word *barometer* (previously called "Torricellian tube" or "weather glass"). In his experiments, Boyle was accompanied by his laboratory assistant, Robert Hooke (1635-1703).

Boyle and Hooke used the barometer in many investigations of atmospheric phenomena. Hooke improved upon von Guericke's pump by making it capable of removing air directly from a vessel. In experiments which followed, Boyle and Hooke demonstrated that in a vacuum, a feather and a coin fell at exactly the same rate, and that the sound of a bell could not be heard. They also discovered that objects would not burn in a vacuum, animals could not live in it, and air was necessary for both combustion and respiration. Nevertheless, in spite of all these experiments, Boyle still questioned the existence of a vacuum.

One of the most noticeable effects discovered during the experimentation of air-pumps was called the *spring of the air*. One might test this fact by using a bicycle pump. When one pushes in a piston, it jumps back. In 1661, Boyle and Hooke performed a related experiment in which they used a thick glass tube, shaped like the letter J, as shown in the accompanying figure. The tube was sealed on the shorter side. They filled the tube with mercury in such a way that there was a certain volume of air in the sealed part. The level of mercury in the longer part was higher than in the other one. In such a device they were able to simultaneously measure pressure and volume. From a series of 44 meticulous measurements, they derived a conclusion that is now called *Boyle's Law*. The law states that at constant temperature,

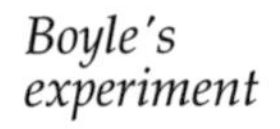

Boyle's experiment

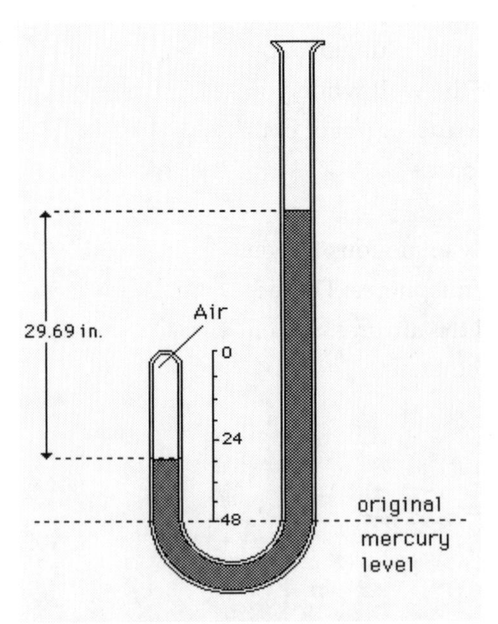

the product of the air pressure P and the air volume V remains constant: P V = constant.

Boyle published this result in 1662 in the Appendix to the second edition of his book, *New Experiments, Physico-Mechanical, Touching the Spring of Air*. Some of his original results are listed in the table below:

Readings at shorter leg of J-tube, proportional to volume V of enclosed air (arbitrary units)	Differences in inches of Hg in short and long legs of J-tube	Pressure P applied to air in inches Hg (atmospheric pressure 29.1 in. added to values in col. 2)	Product PV of values in cols. 1 and 3
48	0	29.1	1397
40	6.2	35.3	1412
32	15.1	44.2	1414
24	29.7	58.8	1411
16	58.1	87.2	1395
12	88.4	117.5	1410
			Mean = 1406.5

The table indicates that the product of pressure and volume is nearly constant. The largest difference from the mean is 11.5 (in the 5th line) which is 0.8% of the mean value.

A French priest, Edme Mariotte (1620-1684) independently discovered the same law. Therefore, some also call it the Boyle-Mariotte law. Thomas Andrew (1813-1885) was the first to observe that below a certain temperature (called *critical temperature*) carbon dioxide does not follow Boyle's law. Later, E. Amget investigated this problem using a long vertical pipe placed in a coal mine in order to obtain pressures higher than in the atmosphere. He concluded that because each gas molecule has its individual volume, the space available for molecular motion must be smaller than a volume V occupied by air. As a result, he found that the equation, p (V - a) = constant (where a is a parameter), agrees more readily with observations for small values of V than the Boyle-Mariotte law.

Robert Boyle (1625-1691)

Robert Boyle was born on 25 January 1625 in Lismore, Ireland. He was the youngest of 14 children in the family of Richard Boyle, the first Earl of Cork and a great Elizabethan adventurer. Robert's first scientific interest was chemistry, However, he published books on heat, thermometry and color. Boyle was one of the founders of the Royal Society. In 1662-1663 Boyle conducted a long philosophical dispute with Spinoza on the question of whether experimentation could provide proof. In contradiction to Spinoza, Boyle argued that experiment was an essential ingredient of proof. He also wrote one of the first English historical novels, *The Martyrdom of Theodora*, about the conflict between love and religious duties. He never married. Boyle died on 30 December 1691 in London.

In 1801, the English chemist, John Dalton (1766-1844) formulated a law which stated that each component of a mixture of gases exerts the same pressure as it alone occupied the whole volume of mixture at the same temperature. Consequently, the total pressure of a mixture of gases equals the sum of the pressures exerted by each constituent gas.

PRESSURE UNITS. *Atmospheric pressure* P is defined as a ratio of the weight F of a column of air and a surface area A below this column. In mathematical terms this definition can be expressed as $P = F/A$. A unit of weight (force; see Chapter 7) is 1 newton (1 N) defined as 1 kg m s^{-2}. A unit of the surface area is 1 m^2. The basic unit of pressure is pascal (Pa): 1 pascal = 1 newton/m^2 = 1 kg m^{-1} s^{-2} There are also other related units of pressure: *bars (b), millibars (mb), millimeters of mercury (mm Hg),* and *atmospheres (atm)*:

1 bar = 1000 millibars = 100,000 pascals
1 millibar = 100 pascals = 1 hectopascal

1 millimeter Hg = 1.3332 millibars = 133.32 pascals
1 atmosphere = 1.01325 bars = 1013.25 millibars =
= 760 millimeters of Hg = 101,325 pascals.

3.6. ATMOSPHERIC PRESSURE

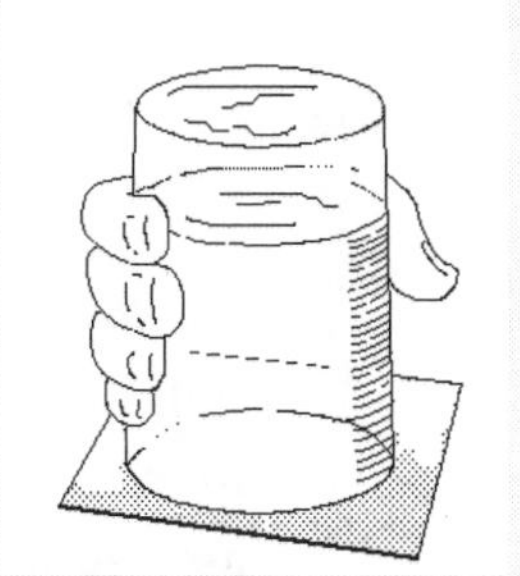

Materials: glass, water, piece of cardboard.

Procedure: Fill a glass with water. Place a piece of cardboard on top of the glass. Cover the glass firmly with the palm of your hand and quickly invert. Remove the palm of your hand carefully from below the cardboard. The cardboard will remain in place.

Explanation: An increase in volume obtained from the bulge in the cardboard causes the necessary pressure drop and allows water to remain in the glass. The experiment works just as well when the glass is full, half full, or almost empty. But if the cardboard is replaced with a piece of aluminum sheet metal, the water will remain in the glass only if the glass is full. This is caused by the difference in flexibility between the cardboard and the aluminum sheet. The cardboard bulges more easily than the aluminum sheet and thus causes a greater change in the volume of air in the glass. When an aluminum sheet is used, the increase in air volume is caused by the slight descent of the water from the glass.

Comments: (i) The experiment also works when a few pinholes are made in the cardboard. This is because surface tension contributes to retaining water in the glass.
(ii) Imagine that the glass filled with water and supported from below only by a piece of cardboard and atmospheric pressure is hung on a spring scale. Will the reading on the scale be smaller than the weight of the glass, cardboard and water combined?
(Answer: It will be equal.)

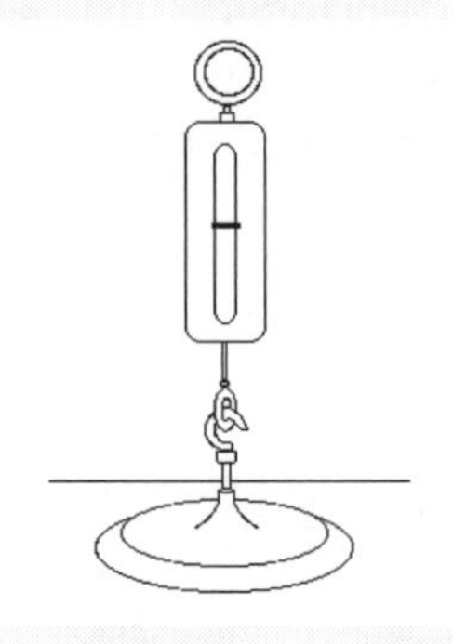

3.7. PRESSURE AND FORCE

Materials: rubber suction cup with attached hook, spring balance, squared paper, plate glass.

Procedure: Attach a rubber suction cup with a hook to a surface. The force required to pull a suction cup away from a smooth surface can be found by using a spring balance. The area on which the atmosphere is pressing on the cup can be measured by using squared paper. The ratio of both values is the atmospheric pressure (since the weight of the suction cup is very small).

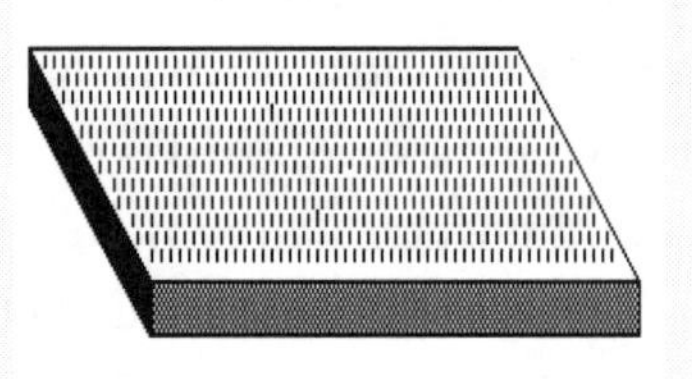

3.8. PRESSURE AND AREA

Materials: 1 x 0.5 m wooden board, 50 nails.

Procedure: Pound 50 nails into the board. The nails should be spaced approximately 1 cm apart. You will be able to lie down on the board and even have weights gently piled upon you.

Comment: This experiment demonstrates that pressure is inversely proportional to area. The larger the area, the smaller the pressure. One can lie down on 50 nails, but one would be unable to safely lie down on only one nail.

The atmospheric pressure on the sea surface is about 76 centimeters of mercury, but it can vary by a few centimeters in either direction. Standard atmospheric pressure at the Earth's surface is often referred to as 1 atmosphere (1 atm). One can easily calculate that: 76 cm of Hg = 1013.25 mb = 101.3 Pa = 1 atm.

Weather facts. The highest sea level pressure, 1083.8 mb, was observed on December 31, 1968 in Agata (Siberia, Russia). The lowest pressure, 870 mb, was detected on October 12, 1979 on Typhoon Tip Island (Pacific Ocean).

ALTITUDE EFFECTS. As seen in the Figure below, the atmospheric pressure rapidly decreases with height. As a result, the environment in the upper atmosphere is very hostile to humans.

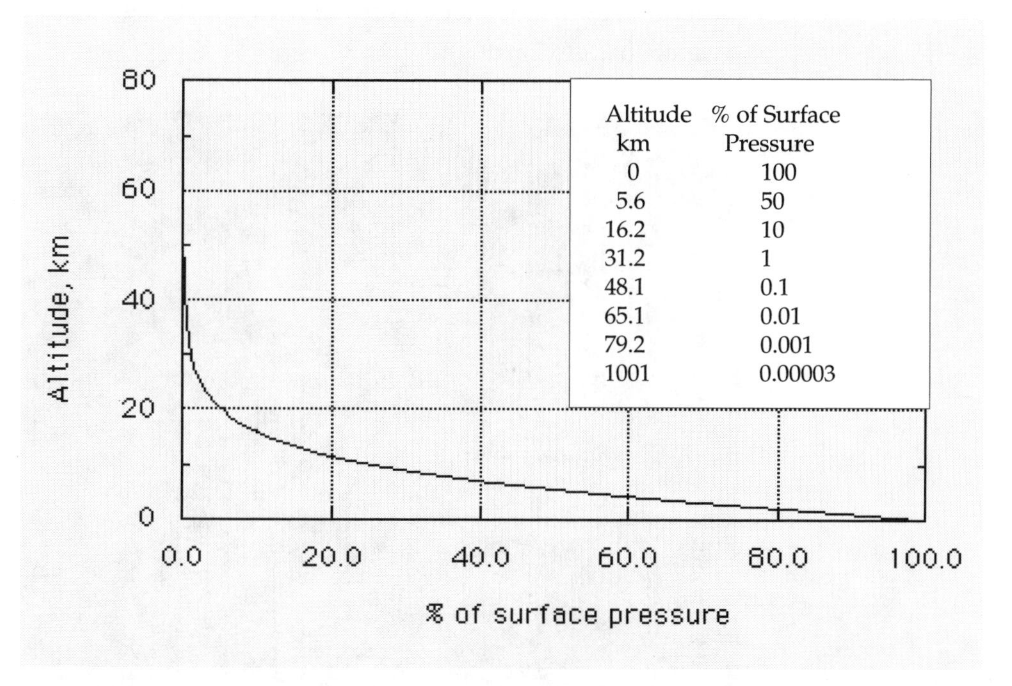

Pressure distribution in the atmosphere

For instance, only with great difficulty are people able to climb summits of mountains higher than 5,000 m. Above this height, every step grows more and more arduous. Respiration becomes more and more difficult. The heart palpitates and the pulse beats more rapidly. Above 12,000 m breathing becomes impossible and life can no longer be sustained.

These facts were experienced by two adventurous English balloonists, James Glaisher and Tracey Coxwell. Their flight on 5 September 1862 initially went smoothly. Problems first arose at an elevation of 6,000 m, where the temperature had fallen to 24° below freezing. At the altitude of about 9,500 m, Glaisher's eyesight began to fail, and he could see neither his watch nor his hands. At 12,000 m he lost consciousness. At the same time his companion, Coxwell, tried to open the gas valve of the balloon to bring it down. He was unable to accomplish the task because his hands were frozen. Nevertheless, he seized the valve's cord with his teeth, and dipped his head until the balloon took a turn downwards, bringing both aviators back to Earth.

Glashier and Coxwell during their flight

TYPES OF BAROMETERS. There are two basic types of barometers. One is a *mercury barometer* (shown in the accompanying figure), not very different from the instrument constructed originally by Torricelli. It consists of a glass bulb with a neck. The tube is filled with mercury and inverted into a basin containing more mercury. Several design modifications have been made to the original mercury barometer, including those of Nicholas Fortin (1750-1821) and John Datton (1766-1844).

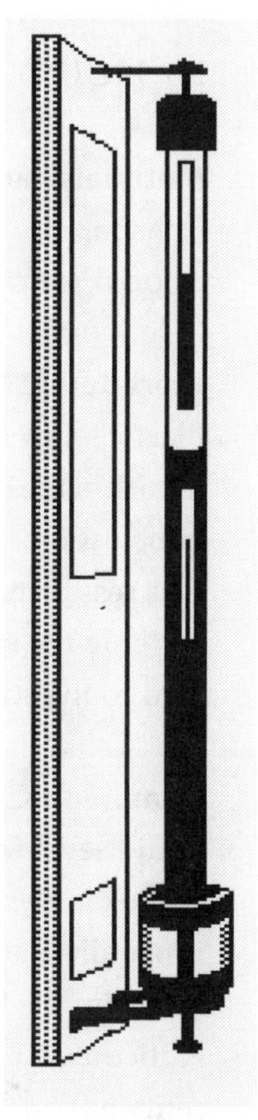

Mercury barometer

Another type of barometer, called an *aneroid*, was initially suggested by Blaise Pascal. A similar design was proposed (however, not constructed) by the famous mathematician Gottfried Wilhelm Leibniz (1646-1716). Leibniz described it in his correspondence to Johan Bernoulli in 1698, and again in 1702. These designs were not implemented until the first practical aneroid was constructed in 1843 by the French engineer, Lucien Vidie (1805-1866).

The aneroid barometer contains no fluid, but consists of a small, flexible metal box called an *aneroid cell*. The pressure inside the cell is slightly lower than the atmospheric pressure. A slight change of atmospheric pressure causes a slight change in the cell's volume, which is registered on a barometric scale. A *barograph* consists of an aneroid cell and a drum equipped with a clock mechanism. The drum turns, moves the chart past the pen arm, and records pressure changes on paper.

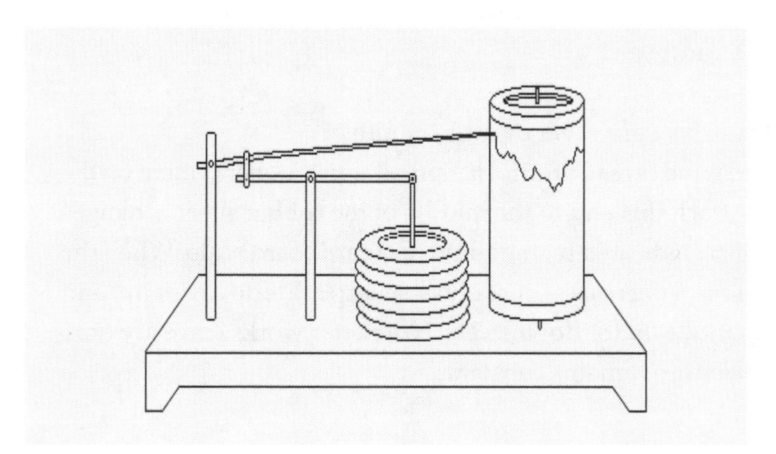

Barograph

3.9. LIQUID BAROMETER

Materials: large bottle, one-hole stopper to fit the opening, transparent plastic straw, about twice as long as the bottle, colored water, white index card.

Procedure: Fill a bottle partly with colored water. Insert the plastic straw through the stopper. Place the stopper into the mouth of the bottle. The lower end of the plastic straw should be below the surface of the water. Blow some air into the tube. As a result, the water will rise above the stopper. The colored water in the straw will rise or sink depending on the air pressure. Attach the card to the straw and mark water level changes.

Comment: Changes of the atmospheric pressure will cause a variation of the liquid level in the straw. The increased pressure will push the level down, while the decreased pressure will force the liquid to rise. Greater air pressure is usually a sign of approaching clear weather while decreasing air pressure indicates the likehood of stormy weather and the possibility of precipitation. Notice also that this barometer works correctly only when the air temperature remains constant.

3.10. ANEROID BAROMETER

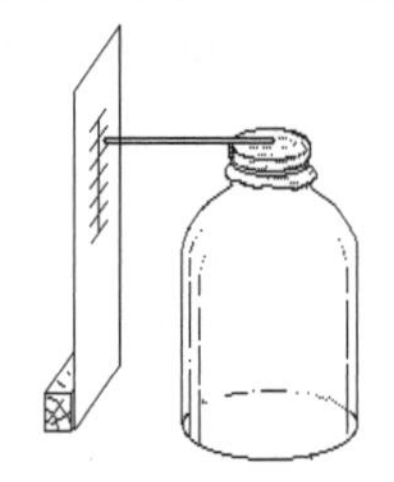

Materials: rubber balloon, glass jar, rubber cement, soda straw, thread, cardboard.

Procedure: Stretch a sheet of rubber over the mouth of a small glass jar and wind thread around it. Spread some rubber cement on the end of the straw. Attach this end to the middle of the rubber sheet. Changes in air pressure will be indicated by the straw on a cardboard scale. When the air pressure increases (or decreases), the rubber sheet pushes down (or up) and forces the straw to move up (or down). The barometer works correctly only when the air temperature remains constant.

Genius is 1 % inspiration and 99 % perspiration. (T.A. Edison)

CHAPTER FOUR
HEAT

Invention of the Thermometer, Thermometric Scales, Gas Laws, Boiling point, Adiabatic process, Heat and Temperature, Heat and Energy, Heat Transfer, Thermodynamics, The Seasons, Temperature variations.

INVENTION OF THE THERMOMETER. *Temperature* characterizes "hotness" and "coldness", and the thermometer is an instrument measuring temperature. A thermometer operates on the principle that heated objects expand while cooled objects contract. The ancient Greeks were already aware of this phenomenon.

About 62 A.D., Heron of Alexandria published *Pneumatica.* In it, he described his famous "temple doors", which opened when an altar fire was burning, and closed again when the fire was extinguished. The device was ingenious (see the figure on the next page). The altar fire caused air to expand in barrel **B**. Expanding air forced some water through siphon **S** into the hanging vessel **V**. The weight of the water in vessel **V** caused it to move down, pull the chains, and thus open the doors. The process was reversed when the fire was extinguished.

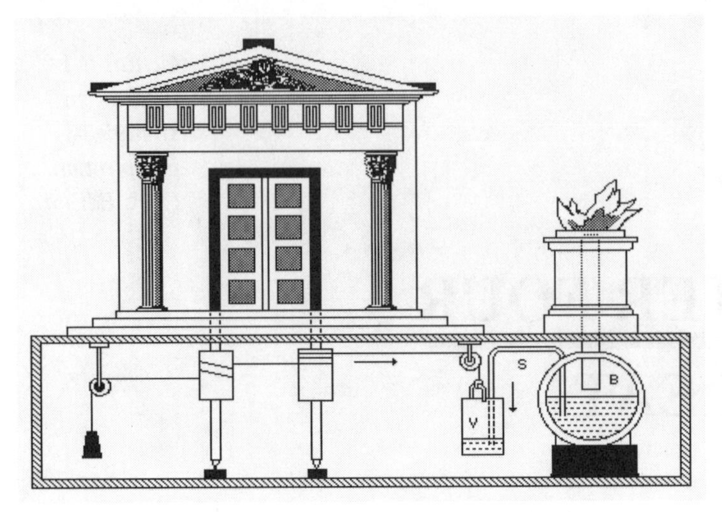

Heron's Temple door

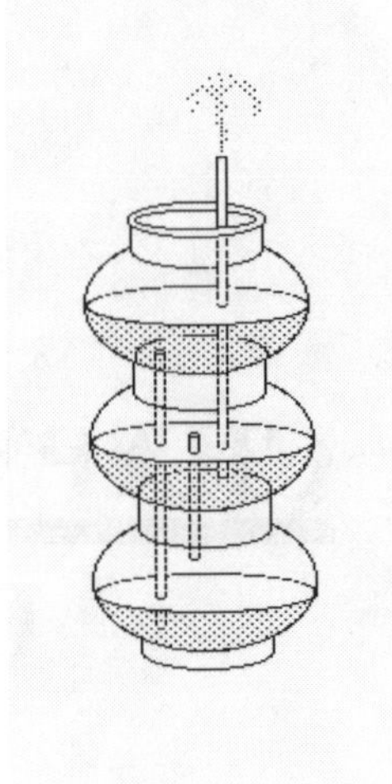

Heron's fountain

In *Pneumatica,* Heron described another clever device. It was a fountain (see the figure above and Exp. 7.2) in which water apparently rose higher than the free surface in its reservoir. Heron's work, translated into Latin in the sixteenth century, caught the attention of many European scholars.

Galileo's thermoscope

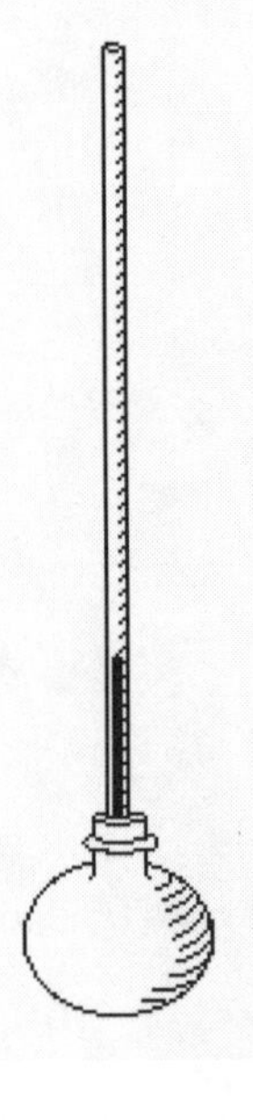

According to Vincenzo Viviani, Galileo's disciple and biographer, Galileo Galilei applied the knowledge of thermal properties of bodies to measure temperature. In about 1593, Galileo allegedly constructed the predecessor of the thermometer. The instrument consisted of a small glass vessel, fitted with a glass tube the width of a straw. The tube was partly filled with water. The expansion of the water in the tube was a measure of heat or cold in the bulb. Galileo also experimented with other thermometric liquids such as salt water, red wine or alcohol. Galileo's thermometer did not have a scale.

The name of the instrument, *thermoscope,* was coined by Giusseppe Biancani in about 1617. Later, when a scale was attached to the instrument, its name was changed to *thermometer.* The word *thermometer* was first used in a book written by the Jesuit priest, Leurechon, in 1624.

Word of Galileo's new instrument spread quickly throughout Europe and many copies were built with numerous improvements. The thermometer reached Poland in 1657, France in 1658, and England in 1661. The invention of barometers revealed that thermometers were subject to changes of atmospheric pressure. Ferdinand II (Medici), Galileo's patron, invented the first sealed thermometer, which was independent of pressure variations.

Ferdinand II de Medici (1620-1670)

Ferdinand II, the son of the Grand Duke Cosimo II, was the ruler of Tuscany. His family, the Medici, had controlled Florence almost continually since the 1420's. Ferdinand II was not only a sponsor of science, but also a scientific hobbyist. He supported Florence's Academy of Experiments (Accademia del Cimento, founded by his brother, Prince Leopold, in 1657, and disbanded in 1667 as a result of Church intervention) and established the first meteorological observation network, with stations in seven Italian cities: Florence, Pisa, Vallombroza, Curtigliano, Bologna, Parma, and Milan. Later the network expanded and included the European cities of Warsaw, Paris, Osnabruck, and Innsbruck. Observations of temperature, pressure, humidity, wind direction, and the condition of the sky were prepared on special forms and sent to Florence for comparison. The figure below shows Ferdinand II participating in an attempt (1657) to determine if the coldness of ice could be reflected by a mirror (it could not).

4.1. WATER THERMOMETER

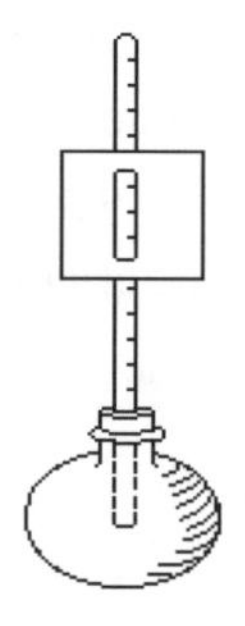

Materials: bottle, one-hole stopper, long tube or straw, thick board to mark a scale.

Procedure: Place a stopper with a long tube or straw into a bottle. Fill the bottle with water so it extends to 1/3 the length of the straw above the stopper. To test the instrument, first place the bottle in a container of hot water. The water in the bottle will expand and force the level in the straw to rise. Then place the bottle in a container of ice. The water will contract and force the level in the straw to drop.

4.2. AIR THERMOMETER

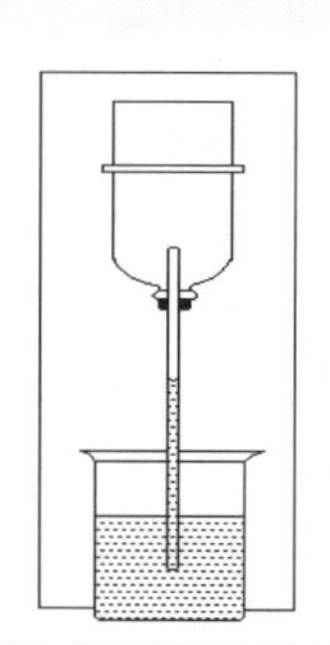

Materials: bottle, one-hole stopper, long glass tube, thick board, jar of water colored with ink.

Procedure: Invert a bottle and attach it to a board. Insert a stopper with a long tube into the bottle. Place the lower end of the tube in a jar of water colored with ink. Warm the bottle with a sponge soaked in hot water. After a few minutes, some air bubbles will move out of the tube into the liquid in the jar. At the same time, a small column of liquid will rise in the tube. The air inside the bottle acts as a sensor. It contracts or expands depending on temperature, allowing the liquid in the tube to rise or drop.

Comment: Notice that this instrument works well only when the atmospheric pressure is constant.

4.3. THERMAL SENSITIVITY

Materials: three containers, water at three different temperatures: hot, lukewarm, and cold.

Procedure: Fill containers to three-quarters capacity: the first one with hot water, the second one with ice water, and the third one with water at room temperature. Place both of your hands, for about half a minute, in the lukewarm water. Next, place your left hand in the hot water, and your right hand in the ice water for about a minute. Quickly dry your hands and put them back into the room temperature water. The water will feel warm to the hand that has been in the cold water, but quite cool to the hand that has been in the hot water.

Comment: Human sensitivity to temperature is very subjective. The same air temperature can be sensed differently by the same person on different occasions. The reason for this is related to the way our skin exchanges heat with the environment. Meteorologists call the temperature which is sensed by a human body the *sensible temperature*. The sensible temperature depends on wind speed as well as humidity and can be evaluated from empirical charts. Generally, windy and humid weather is perceived as chilly, while calm and humid weather is sensed as very hot and muggy.

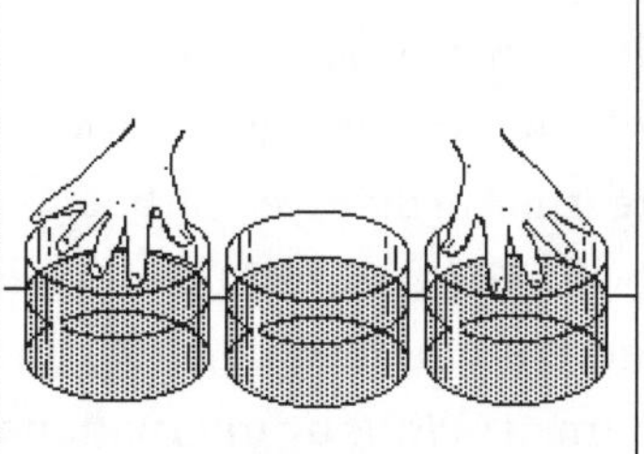

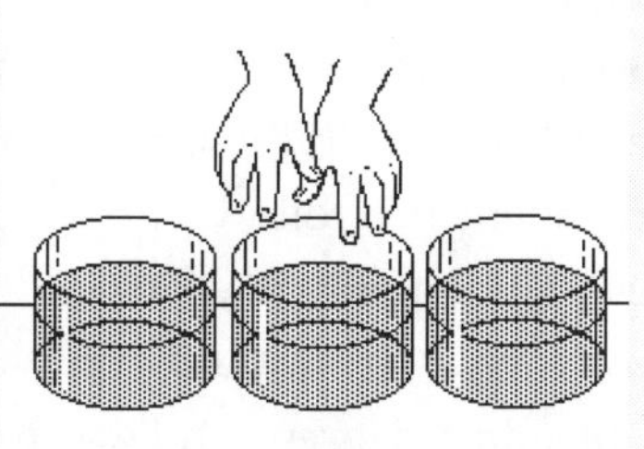

THERMOMETRIC SCALES. Various types of thermometers built throughout Europe employed various temperature scales. Two Englishmen, Robert Boyle (1627-1691) and Robert Hooke (1635-1703), proposed using the melting-point of ice as a fixed point to scale thermometers. In 1665, Christian Huygens (1629-1695) of the Netherlands, suggested using two fixed points.

Gabriel Daniel Fahrenheit (1686-1736) was a German who was born in Gdansk (a city within Polish borders until 1793, and again after 1945). He spent most of his life in the Netherlands, where he designed a thermometric scale based on three fixed points (following a temperature scale used earlier by the Danish astronomer, Olaus Rømer). One point on the scale was marked zero, and was found by placing a thermometer in a mixture of ice, water and sea-salt. The second point on the scale was defined by the temperature of water and ice, and was marked 32. The third point on the scale indicated the temperature of the human body, and was marked 96. As a thermometric liquid, Fahrenheit initially used alcohol, but later (about 1717) switched to mercury. Mercury had certain advantages: it was opaque, remained liquid over a wide range of temperatures, did not evaporate at ordinary temperatures, and was easily obtained in its pure state. A few years after Fahrenheit's death, it became common practice to consider both the freezing point (32°F) and the boiling point (212°F) of water at sea level as fixed points on the Fahrenheit scale.

In 1742, a professor of astronomy at the University of Uppsala in Sweden, Anders Celsius (1701-1744), proposed another thermometer scale. The Celsius scale was defined by two points: at the temperatures of melting ice and boiling water. Originally, Celsius marked the boiling water point as zero, and the ice melting point as 100 (to avoid the minus sign), but later changed it for the sake of convenience, following the suggestion of the Swedish botanist Linnaeus. Until nearly 50 years ago, the unit on the Celsius scale was called *centigrade* (from the Latin word *centum* meaning 100, and *gradus* meaning degree), and was renamed *Celsius* by international agreement to honor the Swedish scientist. The Celsius scale is presently used around the world, except in the United States, where the Fahrenheit scale is preferred by the public. In comparing both scales, temperatures at the two reference points might be considered:

	Celsius	Fahrenheit
freezing water point	0	32
boiling water point	100	212

To find a mathematical relation between the temperatures on both scales, assume that the temperature on the Celsius scale is indicated by t_c, and the temperature on the Fahrenheit scale is indicated by t_f. For example, at the freezing water point $t_c = 0$, and $t_f = 32$. Notice, that at this particular point, $t_c = t_f - 32$. However, this relation is invalid at the boiling point of water, where $t_c = 100$ and $t_f = 212$, because $t_f - 32 = 180$. To obtain the appropriate result, the formula must be divided by 180 and multiplied by 100:

$$t_c = \frac{100}{180}(t_f - 32) = \frac{5}{9}(t_f - 32)$$

This new formula is valid at both (and all other) points. In order to find a temperature on the Celsius scale, providing that the temperature on the Fahrenheit scale is known, you can use the following equation: $t_f = 9/5\ t_c + 32$.

Cricket chirps. Some entomologists (scientists investigating insects) argue that the temperature of the air can be estimated by counting cricket chirps. Crickets are cold-blooded creatures and their activity (chirping) depends on air temperature.

GAS LAWS. In 1661, Robert Boyle and Robert Hooke performed an experiment from which they concluded that if temperature remains constant, the product of the air pressure and volume would also remain constant: $p\,V$ = constant (see also the Section on page 60). Soon after, other important gas relationships were discovered by experimenting with thermometers and barometers.

Daniel Bernoulli (1700-1782) realized that the effect of temperature on gas pressure could be explained by assuming that heat causes the molecules in gas to speed up. As molecules speed up, the harder they hit the walls of the container, and thus pressure rises.

Jacques Charles (1746-1823), a professor at the University of Paris, discovered that gases heated at a constant volume increase their pressure (say, from p_o to p) by $^1/_{273.15}$ part of the initial pressure p_o for each degree Celsius:

$$p - p_o = \frac{p_o}{273.15}\,t$$

where p_o is the initial pressure at $t = 0^\circ$ C.

Charles did not publish any notes about his discovery. It was not until 1802 that Joseph Gay-Lussac (1778-1850) published papers that included this law. Gay-Lussac also obtained a similar equation involving volume. He discovered that gases heated at a constant pressure increased their volume (say, from V_o to V) by $^1/_{273.15}$ part of the initial volume V_o for each degree Celsius:

$$V - V_o = \frac{V_o}{273.15}\,t$$

where V_o is a initial volume at $t = 0^\circ$ C.

The above equation indicates that if a quantity of gas is cooled to $t = -273.15\,^\circ$C, it will be squeezed to a zero volume ($V = 0$). This is supposed to happen at a point known the *absolute zero.* A scale of temperatures counted from that point is called *an absolute scale*. The absolute scale of temperatures was introduced by the Englishman, William Thomson (Lord Kelvin, 1824-1907), and is defined as:

$$T = t_c + 273.15$$

where T indicates temperature on the absolute scale expressed in kelvins (K) and t_c is temperature in degrees Celsius (° C). The word "absolute" means that temperature on the absolute scale is always a positive number. The absolute scale is used by physicists and engineers.

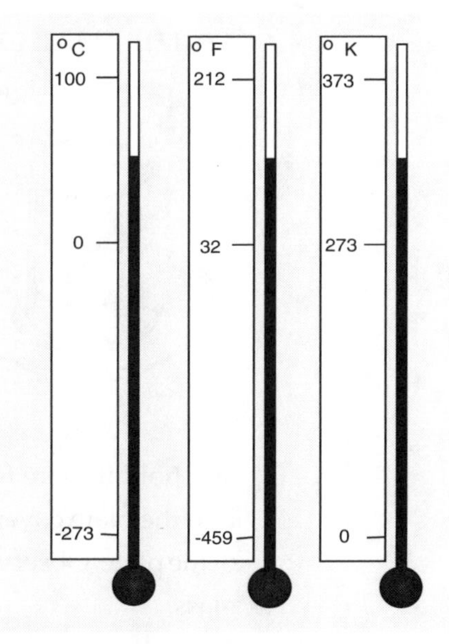

The most common temperature scales: Celsius, Fahrenheit, and Kelvin

4.4. THERMAL CONTRACTION OF AIR

Materials: bottle, small balloon (a hard boiled egg can also be used), pot of hot water.

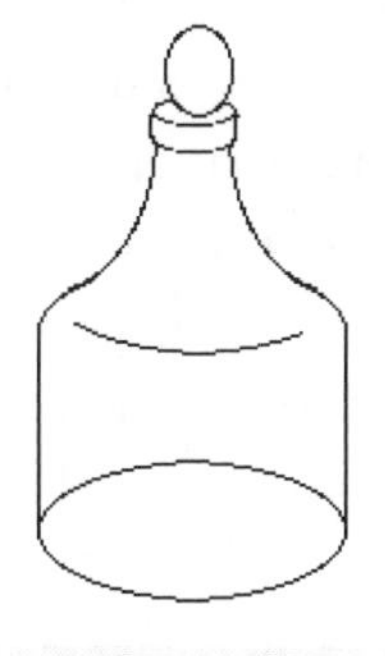

Procedure: Heat air in an empty bottle by placing it in a pot of hot water. Seal the bottle by using a small balloon filled with air. After a while, the balloon will be pushed into the bottle by the atmospheric pressure.

Explanation: When air cools it also contracts and as a result, draws the balloon into the bottle. In order to remove it, blow vigorously into the inverted bottle. This will increase the pressure inside the bottle and push the balloon out.

4.5. THERMAL EXPANSION OF AIR

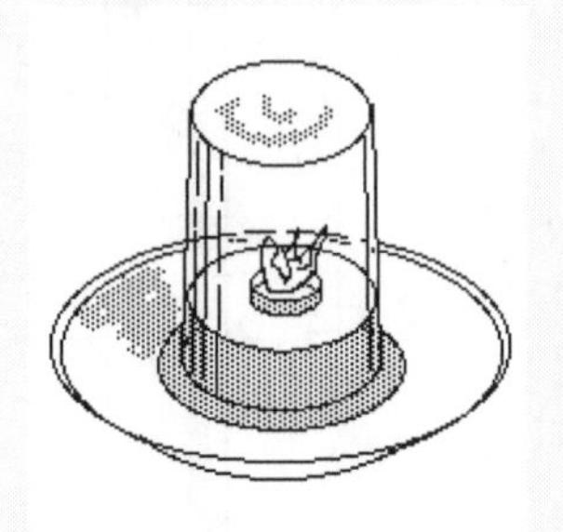

Materials: plate, glass, bottle cap, piece of paper, matches, water.

Procedure: Pour some water into a plate. Place a bottle cap on the water's surface. Put a piece of paper into the cap and light it. Place an inverted glass exactly over the flame and hold it there for about 5 seconds. Slowly lower the glass into the plate covering the cap. Leave the glass on the bottom of the plate. Observe that the water from the plate will enter the glass.

Explanation: This experiment was incorrectly thought of by many to demonstrate that air was made up of approximately 1/5 oxygen. This explanation was wrong because in the place of oxygen, the same amount of other gases, mostly carbon dioxide CO_2, was produced. The real cause of the phenomenon was the thermal expansion and contraction of air. When the paper was burning, hot air in the glass expanded and escaped. Bubbles of escaping air could be observed after the mouth of the glass was put into the water. After the flame was extinguished, the air cooled, contracted and pushed water into the glass.

Comment: This experiment was first performed by John Mayow (1641-1679). Mayow was not in a position to fully explain his observations. However, his work did contribute to further discoveries concerning the nature of air.

It should be mentioned that there is one law which combines all of the individual gas laws and defines the changes of all gas parameters: mass, volume, pressure and temperature. This law is called the *ideal gas law* or the *equation of state* , and has the following form: $pV = mRT$, or $p = \rho RT$, where p is the air pressure (Pa), V is the air volume (m^3), T is the absolute temperature (K), m is the mass of the gas (kg), ρ is the gas density (kg/m^3), and R is the gas constant for dry air, $R = 287$ $m^2s^{-2}K^{-1}$. The behavior of real gases only approximately follows the ideal gas law. This explains why the gases which satisfy this law are called "ideal". For the conditions observed in the atmosphere, the air satisfies the ideal gas law quite well.

BOILING POINT. In 1680, the French physicist, Denis Papin (1647-1714), discovered that when the pressure on a liquid was lowered, its boiling point dropped, and when the pressure on a liquid was increased, its boiling point temperature also increased. For the pressure range 750-1000 mb, the law is illustrated in the following figure:

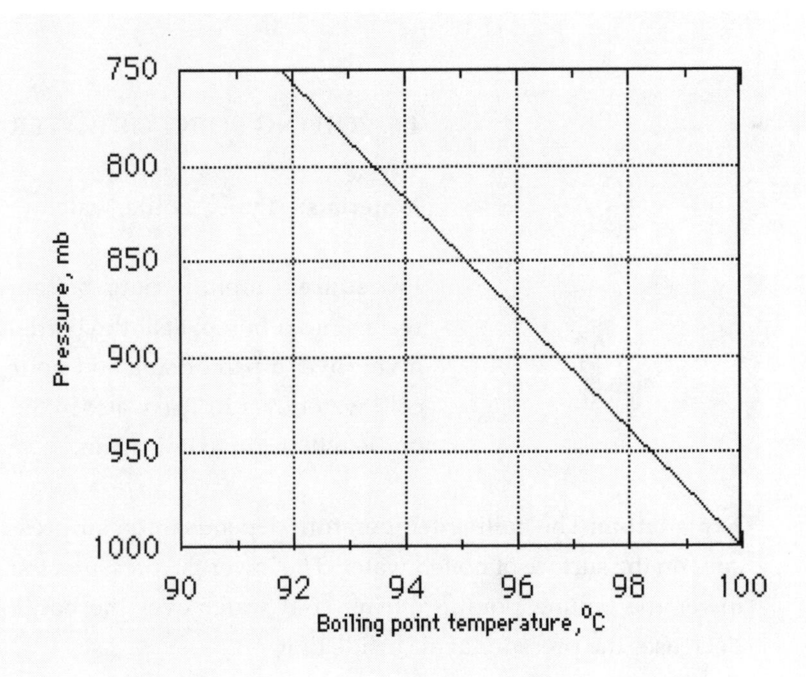

Dependence of boiling point temperature on pressure

Papin's inventions. Denis Papin was a physician, natural philosopher, mechanic, and the inventor of the safety valve and steam engine. He proposed to Louis XIV (1643-1715) to use his engine to pump water for the Versailles fountains, but Coulbert, the pragmatic minister of finance, rejected the offer. Nevertheless, Papin used his engine on a paddle-wheel boat. In 1681, Papin invented the first pressure cooker, called a *digestor*. The digestor (shown in the accompanying figure) increased the boiling temperature and allowed for faster cooking. While visiting England in 1679, Papin prepared a meal in his cooker for the members of the Royal Society. As a result, King Charles II of England (1630-1685, the founder of the Greenwich Observatory in 1675) ordered one digestor for himself. In 1684, Papin moved to London, where he served for three years as a temporary curator of experiments to the Royal Society.

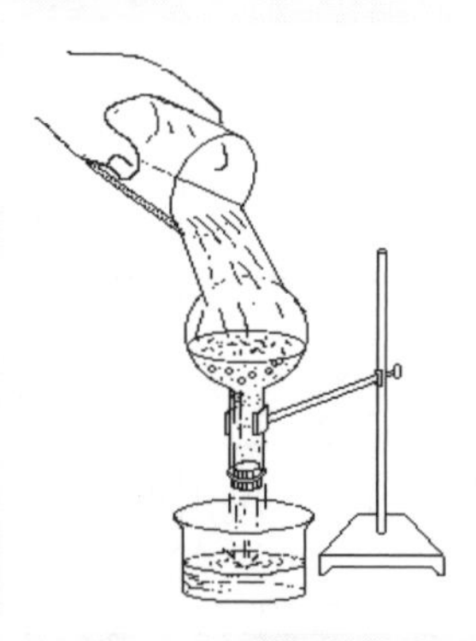

4.6. BOILING POINT OF WATER

Materials: "Pyrex" bottle, water.

Procedure: Carefully pour boiling water into a bottle. Seal the bottle, invert over a pan or sink and pour cold water over it. The water in the bottle will begin to boil again.

Explanation: The boiling temperature depends upon air pressure on the surface of boiled water. The lower the pressure, the lower the boiling point. Pouring cold water over the bottle decreases the pressure of air inside of it.

The pressure 750 mb (in the figure on page 79) is usually observed at roughly 2500 m above sea level. On the top of Mount Everest (the highest point of the Earth, 9000 m above the sea level) the air pressure is about 300 mb, and the boiling point temperature of water is about 70°C. This temperature is not high enough to extract the best flavor from tea leaves. Therefore, do not expect tea to taste good at such an altitude.

ADIABATIC PROCESS. John Dalton (1766-1844) observed that *thermally insulated* air increased its temperature when compressed, and decreased its temperature when allowed to expand. Consequently, he discovered the principle of *adiabatic changes* of temperature. The word *adiabatic* is derived from the Greek word *adiabatos*, meaning "impassible", i.e., occurring without loss or gain of heat. The term *adiabatic process* was first used by a Scottish engineer, William John Rankin (1820-1872).

The adiabatic process follows a more general law, formulated in 1881 by Henri Louis Le Châtelier (1850-1936), and called *Le Châtelier's Principle.* According to this principle: "*if any external changes are imposed upon a system, the system will rearrange to oppose them*".

When a sample of air is compressed, its volume decreases. The air, according to Le Châtelier's Principle, opposes this change by increasing its temperature (increasing temperature causes the air to expand). On the other hand, when a sample of air is decompressed, its volume increases. The air, again, opposes this change by decreasing its temperature (decreasing temperature causes air to contract).

Adiabatic changes accompany the vertical motions of air in the atmosphere. Dry air moving upward expands (because the ambient pressure decreases with height), and in agreement with Le Châtelier's Principle, cools adiabatically. Similarly, the same air moving downward contracts and warms adiabatically. The temperature changes accompanying the vertical motions of air in the atmosphere occur at constant rate of about 1°C per 100 m, called the *dry adiabatic lapse rate.*

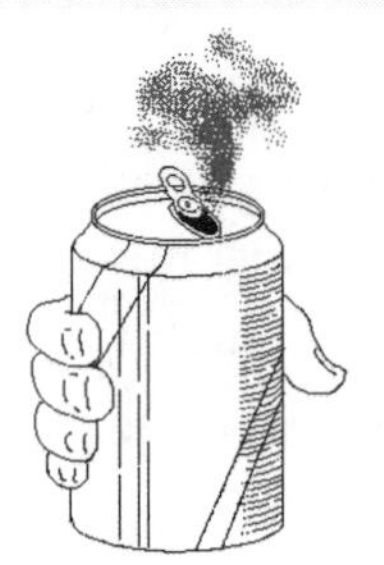

4.7. ADIABATIC PROCESS

Materials: soft drink can.

Procedure: Open the can. Observe the formation of a small cloud above it.

Explanation: As the result of an adiabatic expansion of the gas (CO_2) in the can, the temperature in the bottle (or the can) rapidly decreases. This causes the water vapor in the air to condense which can be seen as a small cloud.

Comment: Similarly, when a piston pump compresses air into a football, gas gains heat when its molecules are forced closer together. A pump frequently gets too hot to touch.
Another experiment demonstrating the adiabatic process can be performed by rapidly expanding a strip of rubber. Notice how the rubber objects contract when heated. In agreement with the *Le Chatelier Principle,* the temperature of the expanded rubber strip will increase (as sensed with one's mouth). Rapid expansion of rubber is equivalent to the adiabatic compression of gases. In both cases there are no heat exchanges between the system and its surroundings.

Weather facts. The hottest country on our globe is Ethiopia, where the mean annual temperature is 39.5°C (103.1°F). The highest observed temperature, +57.8°C (136°F), was registered on September 13, 1922 in El Azizia, Libya. The lowest temperature, -89.6°C (-128.2°F), was detected on July 21, 1983 in Antarctic station Vostok. The coldest city is Vierkhoiansk (Russia), where the mean annual temperature is -68°C (-90.4°F). The largest diurnal temperature change was observed on 23/24 January, 1916, in Browning, Montana, where temperature dropped from +7°C to -49°C during 24 hours. The fastest temperature rise was observed on 26 January, 1933, in South Dakota, where temperature increased from -39°C to 8°C during 3 minutes.

HEAT AND TEMPERATURE. In ordinary language the word *heat* is used as a synonym for *temperature*. However, in science these two words stand for quite different things. *Heat* is a factor which can make things hotter or colder. *Temperature* is a quantity which measures the effect of cooling or heating.

For centuries the nature of heat was one of the great puzzles of the natural world. Since the writings of the Roman author Lucretius (95-55 B.C.), heat was considered to be a type of matter in a fluid form. The term *caloric* (the Latin word *calor* means "heat") was used to designate this particular imponderable fluid, which warmed or cooled. Caloric was thought to be weightless and invisible. Lavoisier even included it in his list of elements.

But the caloric theory could not explain why equal volumes of water and mercury, heated in identical vessels, changed their temperatures differently. Mercury increased its temperature almost twice as much as water. Similarly, when equal weights of iron and water were put side-by-side atop a hot stove, soon the iron became too hot to handle, while the water was still lukewarm. Moreover, it was not understood why the temperature of a mixture of equal quantities of mercury and water, each at different temperatures, was more affected by the initial temperature of water than by the initial temperature of mercury.

In 1764, Joseph Black (1728-1799) noticed the distinction between the temperature, which is the *intensity* of heat measured with a thermometer, and the *quantity* of heat, which is the amount of energy required to heat or cool the body substance. Black provided the following example: "*two pounds of water, equally heated, must contain double the amount of heat that one of them does (though the thermometer applied to them separately, or together, stands at precisely the same point), because it requires double the amount of time to heat two pounds as it does to heat one*".

By mixing equal volumes of water of different temperatures, t_1 and t_2, Black obtained that the mixture had a mean temperature $(t_1+t_2)/2$. He also observed that the amount of heat necessary to produce the same temperature changes of equal masses of various substances was different. As a result, Black defined *specific heat* as the amount of heat necessary to change the temperature of one gram of a substance by one degree Celsius. Based on this definition, a unit of heat could be defined and named a *calorie. A calorie* is the amount of heat necessary to change the temperature of one gram of water by one degree Celsius, from 14° to 15°C.

4.8. SPECIFIC HEAT

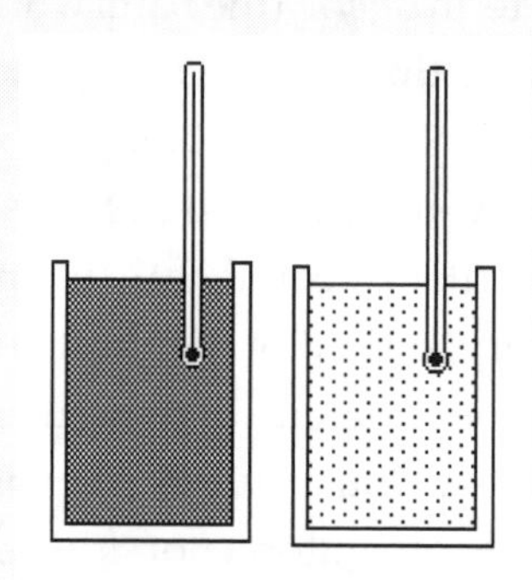

Materials: two thermometers, bucket of soil, bucket of water.

Procedure: Insert one thermometer into a buckets of soil and one into a buckets of water. Read both temperatures. Expose the buckets to the Sun for about an hour. Observe that soil warms more quickly than water. Move both buckets to a cool shaded area. Observe also that soil cools faster than water.

Explanation: Water and soil have different specific heat values so they change their temperatures differently. Moreover, the heating energy is deposited only in a few centimeters of soil and in entire volume of water.

Black also discovered that under certain conditions, heat could be absorbed without temperature changes. In 1761, he showed that when water was cooled to the freezing point, its temperature remained constant, even though the cooling process had been continued. Similarly, when water was heated and its temperature reached the boiling point, it remained constant, even though the heating had been continued. Black concluded that *melting, freezing, evaporation*, and *condensation* were accompanied with the passage of heat, called *latent heat*. The latent (from a Latin word for "hidden") heat was released or absorbed during phase transformations (ice-water-vapor) in such a way that temperature remained constant.

Black also evaluated how much heat had to be given away by water in order to freeze. That amount was 82 calories per gram of water. The amount of heat that had to be used in order to melt ice was also 82 calories per gram of ice. Black's result compares well with the modern value of the latent heat of melting or freezing at 0°C which is 79.9 cal/g. The modern value of the latent heat of evaporation at 0°C is 597.3 cal/g.

HEAT AND ENERGY. Two and a half decades later, in 1798, an American, Benjamin Thompson (Count Rumford, 1753-1814), observed that intense heat was generated by friction when brass cannons were bored with drills. Thompson performed an experiment in an arsenal in Munich by drilling a brass cylinder with an initial temperature of 60°F (15°C). The steel drill, 0.63 of an inch thick and 4 inches long, was pressed against the cylinder with a force of about 10,000 pounds. The drill was also rotated about 32 times per minute, by using the power of two horses.

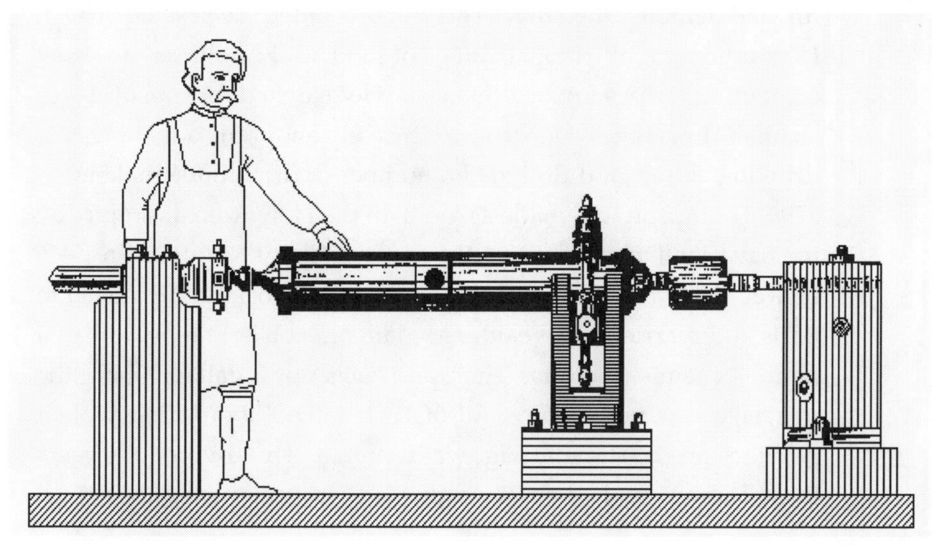

Gun drilling

Benjamin Thompson (Count Rumford) (1753-1814)

Benjamin Thompson was born on 26 March 1753 in Woburn, Massachusetts, into a family of small village farmers. He had little formal education. All that he learned was by self-study with the help of local clergymen. In 1772, he moved to Concord (previously called Rumford), New Hampshire, where he married a wealthy widow, 14 years his senior. They had one child, but separated three years later. During the American War of Independence, after the fall of Boston, Thompson emigrated to Europe. In England, his career progressed rapidly. He became the Undersecretary of State in the Ministry for the Colonies, and was knighted by King George III. Later he moved to Bavaria, where he was nominated Minister of War. During his duty as a military commander, Thompson demonstrated a genius for technological improvements. For example, he made studies of the insulating properties of cloth and fur, and demonstrated that a vacuum prevented the passing of heat. He studied the nutritional values of food, and published an essay containing recipes for healthy meals. He searched for substitutes of alcoholic beverages. He wrote extensively about the advantages of drinking coffee, and designed a number of drip coffee makers. In 1793, Thompson was made a Count of the Holy Roman Empire by the Bavarian Duke, adopting the name of Count Rumford. In 1799, he was elected to the Fellowship of the Royal Society. He moved to Paris and married the wealthy Madame Lavoisier, the widow of a famous chemist. Because Thompson was not a gallant man, the marriage was not happy. All of Paris talked about their violent public quarrels. The marriage was terminated by divorce two years later. Thompson died on 21 August 1814 in Auteuil, France.

After half an hour, and 960 revolutions, the temperature of the cylinder increased to 130°F (50°C). The cylinder and the bored-out scraps were carefully weighed before and after the experiment. The mass of the cylinder was 113.13 pounds. The bored-out scraps weighed 1/948 part of the mass of the cylinder. Rumford found it hard to believe that "caloric" from such a small amount of bored-out scraps could produce such a great temperature increase in the large cylinder.

In a second experiment, Rumford used a brass cylinder drilled under 2 gallons of water with an initial temperature of 60°F (15°C). After an hour, the temperature of the water increased by 47°F. After 2 hours, the temperature of the water was 178°F (75°C). After 2 hours and 20 minutes, the temperature of the water was 200°F (85°C). Finally, after 2.5 hours, to the amusement of spectators, the water started to boil. Rumford calculated that the amount of heat produced by drilling was greater than the amount of heat given off from 10 burning candles, each with a diameter of 3/4 of an inch. Because it seemed that the heat produced by friction was almost inexhaustible, Rumford realized that "caloric" could not be a material substance at all, and therefore he rejected the caloric theory.

In the 1840's, James Prescott Joule (1818-1889) of Manchester, England, studied the conversion of mechanical energy into heat. In one of his experiments (see the accompanying figure), paddles were rotated inside a cast iron vessel filled with water. The paddles were turned by a falling weight, attached to a rope wrapped around a vertical rod and passed over a pulley. The paddles created friction, thus raising the water's temperature. The complex arrangement of paddles ensured that work performed

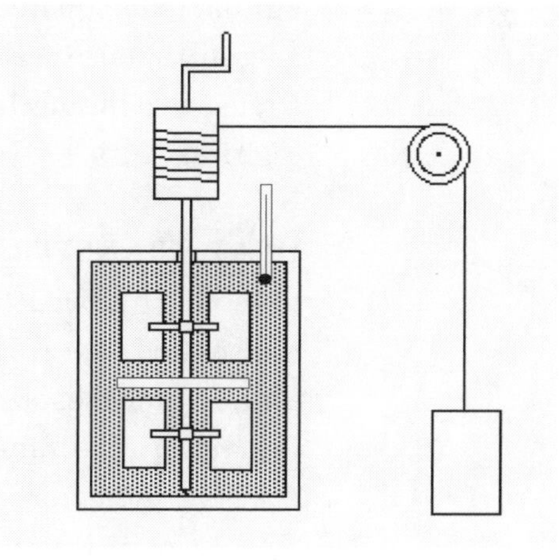

Joule's experiment

by the falling weight was transformed into heat rather than into mechanical energy (fluid motion) again. Joule measured the mechanical energy of the falling weight (the friction arising from the pulley and the rigidity of the string was subtracted) and then measured the rise in temperature. The experiment proved that heat was not a substance (if it were, it could be created or destroyed by mechanical process) but a form of energy.

Ultimate recognition. Joule worked as a brewer by trade. Therefore, his work was first met with scepticism and his paper on heat and energy was rejected by various professional journals. In 1847, Joule presented his results at a public lecture, raising the interests of twenty-three-year-old William Thompson, later known as Lord Kelvin. Due to Thompson's interest, Joule's discovery was finally recognized three years later, in 1850.

Joule was honored by having a unit of energy named after him. A unit of energy (and also of work), joule (1 J), is defined as a product of the unit of mass of a moving object (1 kg) and the squared unit of its velocity (1 m^2/s^2): 1 joule = 1 kg m^2/s^2 (also see Chapter 7). Joule's experiment demonstrated that the mechanical equivalent of heat is: 1 joule = 4.186 calories.

HEAT TRANSFER. Heat can be transferred from one object to another by *conduction*, *convection*, and *radiation*. When heat is transmitted as a result of collisions by rapidly moving molecules within the substance itself, the process is called *conduction*. *Convection* is the transfer of heat by a flow of a heated substance. *Radiation* is the transport of energy by electromagnetic waves.

Count Rumford devised a flow indicator to observe convection currents in liquid. He used a liter glass tube with a long cylindrical neck into which he put about a half teaspoon of yellow amber powder. Then distilled water was mixed with an alkaline solution to make the amber powder buoy up and suspend in the fluid.

All three processes, *conduction*, *convection*, and *radiation*, occur when water is warmed in a pot. First, a thin layer of water on the bottom is heated by conduction. Then the absorbed heat is redistributed in water by convection, and emitted in a form of radiative energy. In addition, some energy is also transferred to the air due to evaporation in the form of the latent heat.

4.9. GREENHOUSE EFFECT

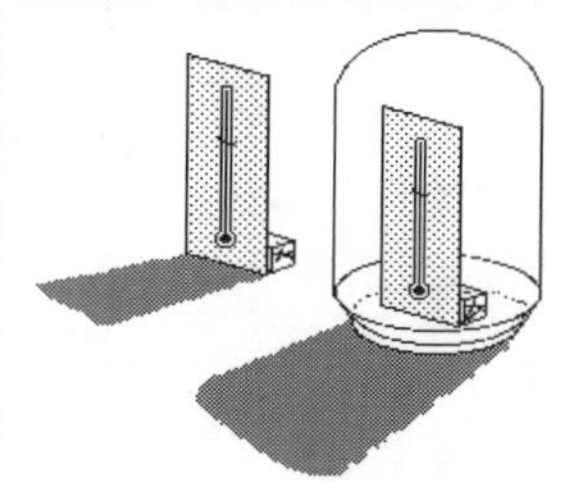

Materials: large thick glass jar, two thermometers, two pieces of cardboard.

Procedure: Position two thermometers upright by resting them against two pieces of cardboard in a sunny and quiet spot. Do not allow the Sun to shine directly on either thermometer. Cover one of the thermometers with a jar, as shown in the figure. Observe that the thermometer in the jar will register a higher temperature than the other one.

Explanation: The primary mechanism for keeping the air in the jar warmer is the suppression of convection and preventing the exchange of air (and heat) between the inside and outside. Therefore, the jar acts as a horticultural "greenhouse".

Comment: In the atmosphere, water vapor, carbon dioxide and other gases absorb the terrestrial radiation, and thus partly prevent it from escaping to space. The surface of the Earth receives energy from the Sun and from the atmosphere, and therefore it is warmer than it would be in the absence of the named gases in the atmosphere. Even though such warming of the Earth's surface is called "the greenhouse effect", its basic cause is only due to atmospheric radiation, and not due to suppressed convection, as in a real greenhouse.

4.10. RADIATION

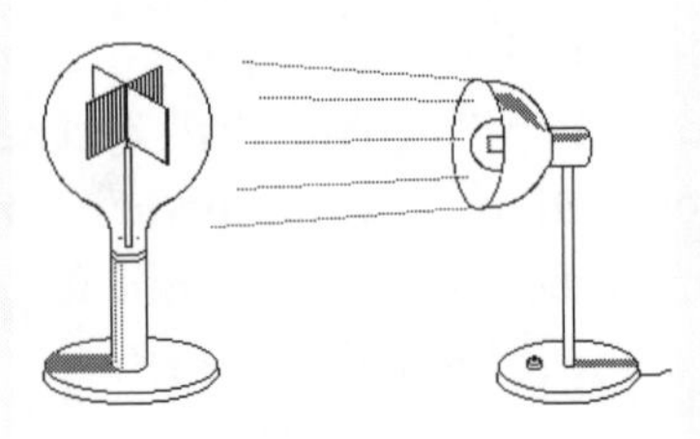

Materials: radiometer (available in toy shops), lamp.

Procedure: Observe the motion of the radiometer.

Explanation: The radiometer is set in motion by radiant heat energy. The black side of each vane absorbs radiant energy and becomes warm. The light side reflects it and stays cool. Gas near the black, warmer sides expands and pushes the vanes into circular motion.

4.11. HEAT CONVECTION

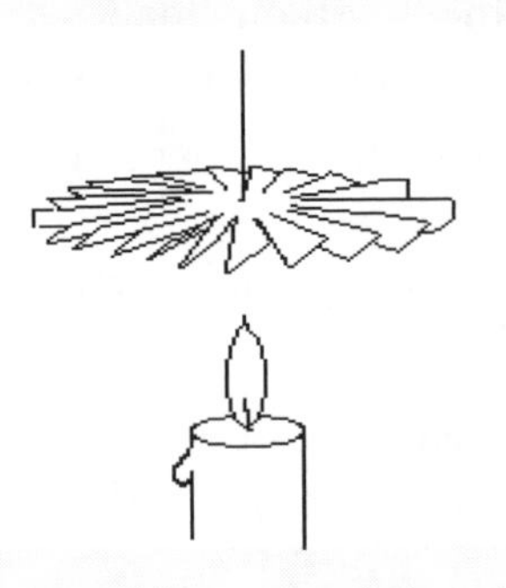

Materials: piece of aluminum foil, wire, candle.

Procedure: Cut a circular disk out of paper or foil. Make a rotor by cutting teeth in the outer edges of the circle. Put angle on "teeth" for best result. Support the rotor on a bent wire. Hold it over the flame of a candle. The rotor will revolve rapidly.

Explanation: The heat of a burning candle decreases the air density. The hot, light air rises and causes the rotor to move.

4.12. HEAT CONDUCTION

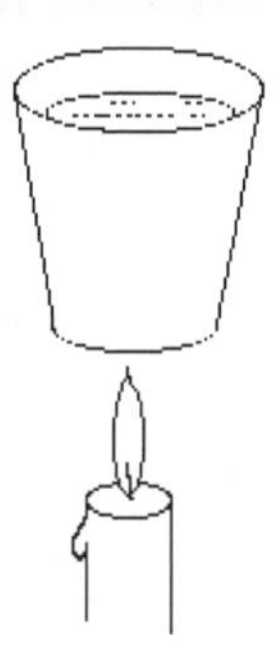

Materials: paper box (or styrofoam cup), water, candle.

Procedure: Fill a paper box about half full of water and place it over a candle or burner. The water can be boiled without burning the paper box.

Explanation: The paper conducts heat from the flame to the water and as a result stays cool. It will not catch fire because the temperature at which the paper ignites is higher than the boiling point of water.

THERMODYNAMICS. *Thermodynamics* is the science of heat and energy transfer. Thermodynamics was developed in the 18th, 19th and early 20th centuries as a result of the exploitation of steam engine technology. However, certain thermodynamic ideas are at least two thousand years old and can be found in the writings of two Greeks who lived in the first century A.D., Philon of Byzantium and Heron of Alexandria.

Philon described an experiment in which air from a heated leaden globe entered a vessel filled with water and produced bubbles. Heron constructed a primitive steam jet-engine and described it in his work *Pneumatica*. His engine, which he called *aeolipile* (from *Aeolus* - "god of winds", and *pila* - "ball"), consisted of a hollow sphere mounted to rotate about a central axis. When

An aeolipile

the vessel was filled with water and heated, steam escaped from the two tubes placed on opposite sides of the vessel. This caused the device to rotate.

Between then and the seventeenth century, very little was accomplished in the field of heat energy. Then, Otto von Guericke (1602-1686) developed the air pump. Denis Papin (1647-1714) and Christian Huygens (1629-1695) designed a steam engine, by utilizing the idea of a piston in a cylinder from von Guericke. They also built a gunpowder engine, which was the prototype of the *combustion engine*.

In England, Thomas Savery (1650-1715) built a steam pump called the "Miner's friend", which was powered by the pressure of steam and by the vacuum obtained when steam condensed. A similar idea was independently used by another Englishman, Thomas Newcomen (1663-1729), who designed an improved version of the steam engine.

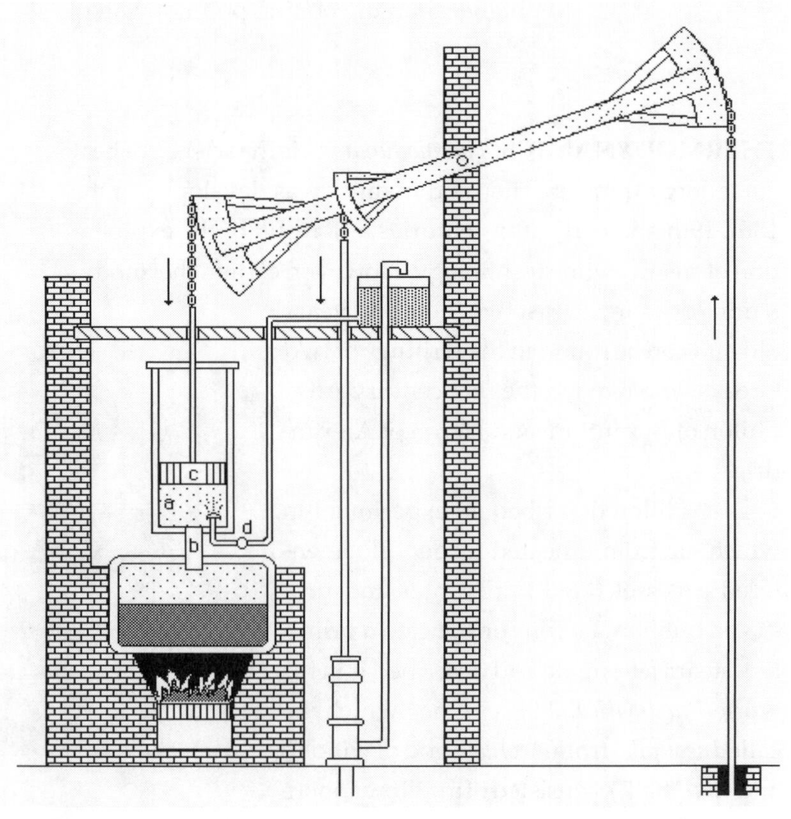

An early steam engine by Newcomen

The way Newcomen's engine worked is shown in the figure on the previous page. Steam was admitted into a cylinder (a), by a tap (b), forcing the piston (c) up. The tap (b) was then closed, and another tap (d) was opened, sprinkling cold water into the cylinder (a). As a result, steam condensed, pressure in the cylinder decreased, and the external atmospheric pressure forced the piston down.

In 1824, a French engineer and physicist, Nicolas Leonard Sadi Carnot (1796-1832), presented the general theory of the steam engine in his pamphlet, *A Reflection on the Motive Power of Heat*. It was about general heat engines. The operation of a general heat engine (the so-called *Carnot's engine*) was reduced to four stages: *isothermal* (at constant temperature) *expansion*, as the steam is introduced into the cylinder, an *adiabatic expansion*, *isothermal compression* in the condenser, and *adiabatic compression*.

Nicolas Leonard Sadi Carnot (1796-1832)

Nicolas Leonard Sadi Carnot was born on 1 June 1796 in the Palace du Petit-Luxembourg in Paris, the son of Lazare Carnot, the Minister of War during the French Revolution (and later during Napoleon's Hundred Days). Nicolas was admitted to the Ecole Polytechnique at age 16. He became a captain in the Corps of Engineers. During his lifetime only one of his manuscripts was published (*Réflections sur la puissance motrice du feu et sur les machines propres 'a développer sette puissance,* in 1824). The manuscript was formally presented to the Académie des Sciences and was favorably received. In 1831, Carnot began investigating the physical properties of gases and the relationship between temperature and pressure. On 24 August 1832, he died of cholera, at age 36. All of his personal belongings, including nearly all of his papers, were burned. Carnot is credited with being one of the founders of thermodynamics.

In his work, Carnot also set forth a few postulates. The first one was the impossibility of *perpetual motion* (motion generated by itself). Using the analogy of a water wheel performing work, Carnot argued that heat flows from a higher temperature to a lower temperature body to generate work. Carnot's ideas were new and difficult, and attracted little attention.

In 1842, the German physician, Julius Robert Mayer (1814-1878), presented his theory, which clarified many concepts. Mayer was a physician on a Dutch sailing ship and for a time practiced medicine on the island of Java. In Java, he observed that the blood in the veins of his Javanese patients exhibited a bright red color, usually found only in highly oxygenated artery blood. In 1840, based on Lavoisier's theory of combustion, he concluded that in hot climates the body required less oxidation to maintain its temperature than would be required in cooler climates. In a paper, *Remarks on the Forces of Inorganic Nature,* published in 1842, he also concluded that the total energy of the world is constant.

Mayer's ordeal. Initially, Mayer's ideas were not appreciated. As a result, he became seriously depressed. In 1848, after the death of his two children, Mayer attempted suicide by throwing himself from the third floor and was confined to a mental institution. Mayer was finally released, but never fully recovered. His achievements were finally recognized in 1871, when he was awarded the prestigious Copley medal of the Royal Society.

In 1850, Rudolf Clausius (1822-1888) of Germany defined *entropy* (from the Greek word *trope,* meaning "transformation"), as the unavailability of energy for further energy transformations (changes into heat or other forms of energy). In the real world there is always energy lost due to such irreversible processes as mixing or friction. Therefore, entropy of real systems continuously increases.

Clausius predicted that a time will come (probably in billion of years) when the temperature will spread uniformly over the entire Universe. During this state, called "*heat death*" of the Universe, there will be no energy available for further transformations. The entropy of the Universe will reach its maximum value.

The main laws of thermodynamics were finally formulated in 1851. The first two were stated by Clausius, and the third one by William Thomson (Lord Kelvin, 1824-1907). Briefly, they are:

(1) *Energy can neither be created nor destroyed. It can only be changed from one form of energy into another,*

(2) *Heat passes from a warmer to a colder body. An engine that would continue in motion for ever ("perpetual motion") cannot be constructed,*

(3) *It is impossible to cool a body below the temperature of the "absolute zero", which is 0 K or -273.15 °C.*

4.13. HEAT ENGINE

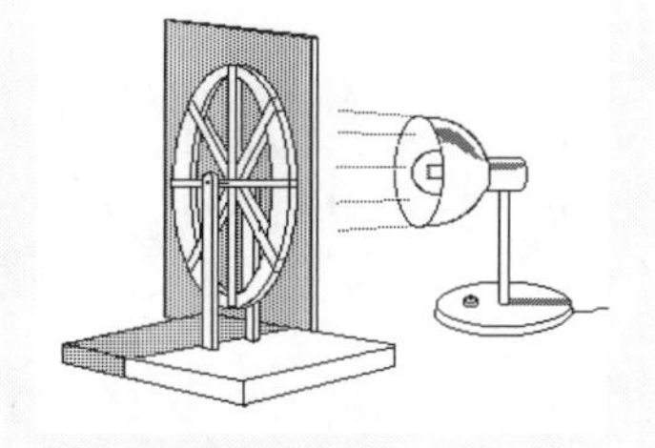

Materials: four rubber bands, sturdy cardboard, bright electric light source.

Procedure: Cut a ring out of the cardboard, with an outer diameter of 10 cm, and an inner diameter of 8 cm. Put the rubber bands around the ring. Pass the pin through the rubber bands in the center of the ring, where they all meet. The pin serves as an axis on the wheel with rubber band spokes. Support the pin on both sides. Focus the light on one side of the wheel, while shading the other side. Observe that the wheel starts to rotate around its axis.

Explanation: Heat causes one of the rubber bands to shrink. This moves the wheel's center of gravity and causes it to turn. As a result, the heated rubber band moves away from the spot where it was heated. It cools and expands. However, another band is now heated, which moves the wheel.

THE SEASONS. The Earth's atmosphere, like a giant heat engine, transforms available energy into the movement of huge masses of air. Practically, all "fuel" for this engine is supplied by the Sun. The contribution of all other sources (e.g., the Earth's interior) is smaller than 0.02%.

The distance between the Earth and the Sun averages about 150 million kilometers. Because the Earth's orbit is not perfectly circular but elliptical, the distance varies during the course of the year. On January 3 of each year, our planet is approximately 147 million kilometers from the Sun, which is closer to the Sun than at any other time. On July 4, about six months later, the Earth is 152 million km from the Sun, farther from the Sun than at any other time. Direct sunlight is only about 7% more intense on January 4 than on July 4. Therefore, the differences in the amount of solar radiation received by the Earth as the result of its elliptical orbit are slight and play only a minor role in producing seasonal temperature variations. But if the distance variation between the Earth and the Sun does not cause the seasons, what does?

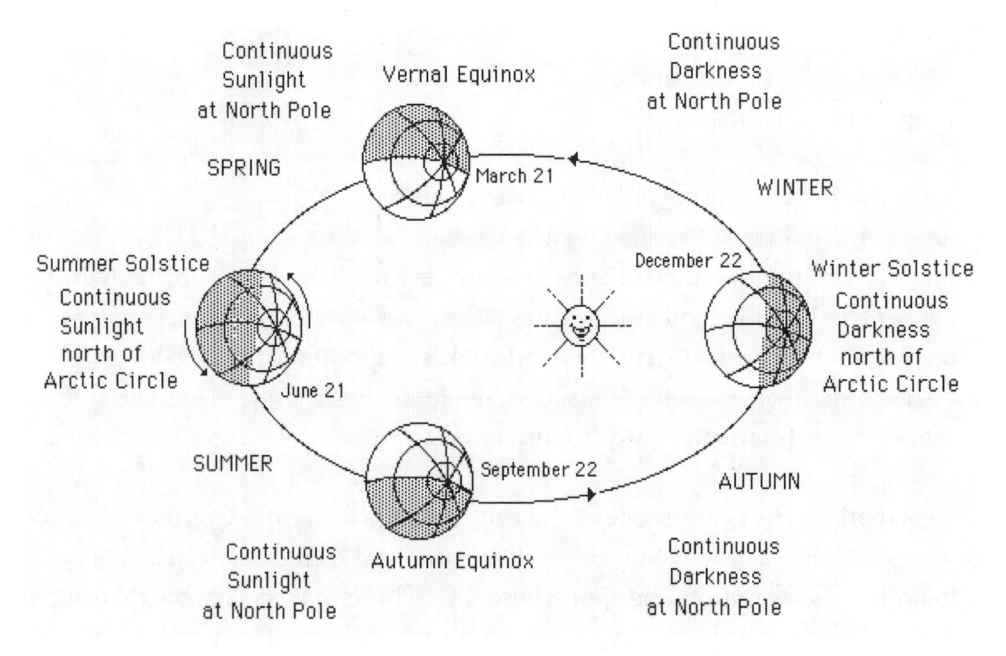

The Earth's orbit

The seasonal variation of the Sun's angle affects the amount of energy received at the Earth's surface in two ways. First, when the Sun is directly overhead, the solar rays are most concentrated on the surface (see Experiment 4.14). The lower the Sun, the more spread out and less intense is the radiation that reaches the surface. Secondly, the angle of the Sun determines the amount of atmospheric depth that a ray must traverse. When the Sun is directly overhead, the rays only pass through a thickness of 1 atmosphere's height, whereas 5° (above horizon) rays travel through a thickness roughly equal to 11 atmospheres. The longer the path, the greater is the chance for absorption and scattering by the atmosphere, which reduces the intensity of the solar radiation at the surface.

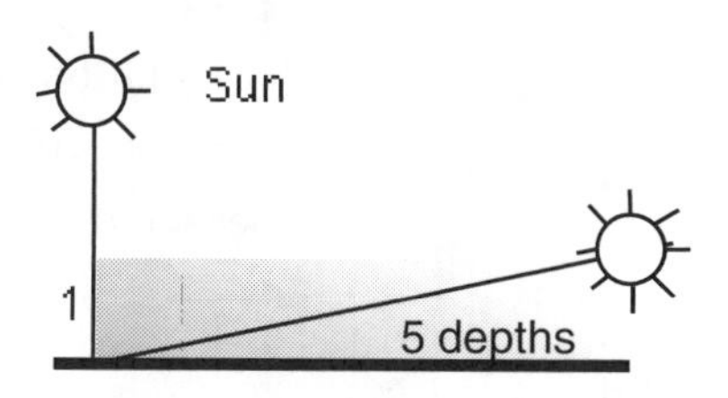

Atmospheric depth

4.14. ANGLE BETWEEN BEAM AND SURFACE

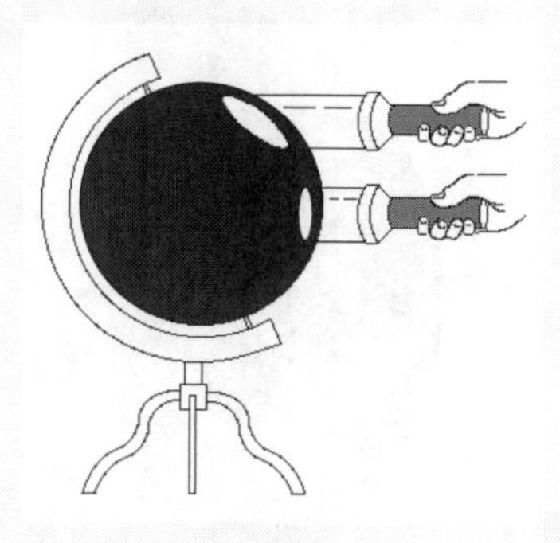

Materials: flashlight, ball or balloon.

Procedure: Direct the flashlight at different angles to the spherical surface. Observe that when the beam perpendicular to the surface, a small intense spot is produced. When the flashlight beam strikes at an oblique angle, the illuminated area is larger and dimmer.

Comment: The relationship between the intensity and angle of incidence can be described as a cosine relationship, also known as Lambert's law. It was first derived by Johann Heinrich Lambert (1728-1777), a German mathematician. Lambert also invented a word *albedo* (in Latin "whiteness") for the fraction of light reflected by a body.

The angle between the beam and the surface changes as a result of the rotation of the Earth around its axis (diurnal variation), and also as a result of the revolution of the Earth around the Sun (annual variation). The Earth's axis is not perpendicular to the plane of its orbit around the Sun, but is tilted at the angle of 23.5°. This angle is called the *inclination* of the axis (see figure below). Because the axis remains pointed toward the same direction in space (toward the North Star at the present epoch), as the Earth journeys around the Sun, the orientation of the Earth's axis to the Sun's rays consistently changes. If the axis were not inclined, there would be no seasonal changes.

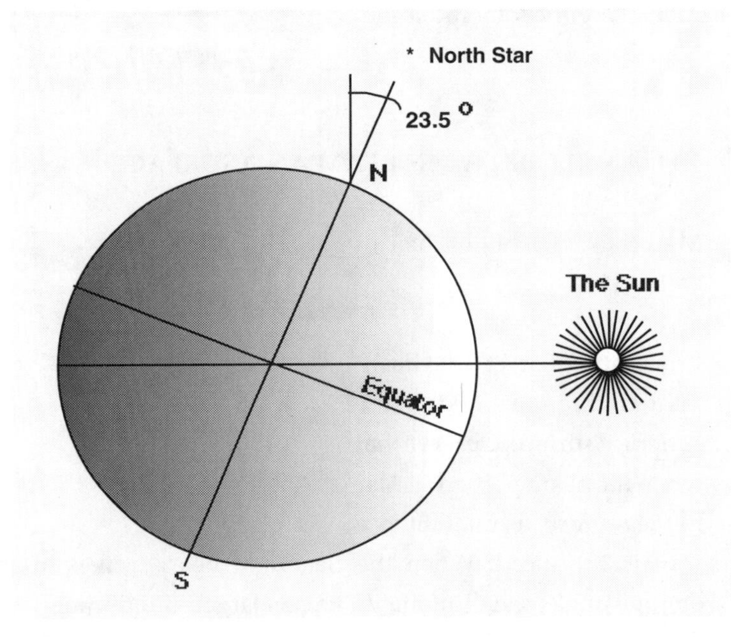

Inclination of the Earth

On one day each year, the axis is such that the Northern Hemisphere slants 23.5° away from the Sun. Six months later, when the Earth has moved to the opposite side of its orbit, the Northern Hemisphere slants 23.5° toward the Sun. On days between these extremes, the Earth's axis is oriented at angles less than 23.5° to the rays of the Sun. These changes cause the "overhead Sun" to make a yearly migration from 23.5° north to 23.5° south of the Equator.

Four days in the year have been given special significance based on the annual migration of the Sun across the sky. On June 21 or 22, the Earth is in a position where the axis in the Northern Hemisphere is tilted 23.5° toward the Sun. At this time, the rays of the Sun are striking vertically at the latitude 23.5° north of the Equator, at an imaginary line called the *Tropic of Cancer*. For people who live in the Northern Hemisphere, this day is known as the *summer solstice* (the day with the longest length of daylight in the Northern Hemisphere). Six months later, on December 21 or 22, the day known as the *winter solstice* (the day with the shortest length of daylight in the Northern Hemisphere), the Earth is in a position where the axis in the Northern Hemisphere is tilted 23.5° away from the Sun. The rays of the Sun are striking vertically at the latitude 23.5° south of the Equator called the *Tropic of Capricorn*.

The *equinoxes* occur midway between the solstices. On these two 12 hour-days, the vertical rays of the Sun are striking at the Equator, and the axis of the Earth is perpendicular to the solar rays. Because the Earth undergoes a change in speed in its orbital motion (in accordance with Kepler's laws), the period between the winter solstice and the spring equinox is approximately 89 days, and 93 days between the spring equinox and the summer solstice. September 22 or 23 is the date of the *autumnal equinox* in the Northern Hemisphere, while March 21 or 22 is the date of the *vernal (spring) equinox*.

TEMPERATURE VARIATIONS. Temperature in the atmosphere significantly varies in time (diurnal and seasonal variations), with height, and from place to place on the Earth. These variations are controlled by several factors such as the Sun's elevation above the horizon, the duration of the daylight, cloudiness, variation of specific heat between land and water, albedo (reflecting ability of the surface with respect to solar radiation), and the influence of ocean currents.

The underlying surface of the Earth has a considerable influence on the air temperature. For example, a cold continental air mass in the winter is quickly heated over the warmer ocean, while cold air from the ocean is warmed when it passes over the warmer land in the summer.

Differences in temperature near the ground are caused by the variation of thermal properties of the underlying surface. For instance, it takes far more heat to raise the temperature of water than it does to raise the temperature of rocks or soil. This is because water has an enormous specific heat. Moreover, the heating energy is deposited only in a few centimeters of soil, while in the oceans it is mixed through the top few meters of water. Consequently, since water covers 61% of the Northern Hemisphere, and 81% of the Southern Hemisphere, there are considerably smaller annual temperature variations in the water-dominated Southern Hemisphere, as compared to the Northern one.

The thermal effects of ocean currents are also significant. The Gulf Stream (the oceanic stream initiated in the Gulf of Mexico and extending toward Europe) keeps wintertime temperatures of western Europe warmer than would be expected for their latitudes. For example, a mean January temperature in Berlin (52°N) is about equal to a mean January temperature in New York which lies 12° (about 1200 km) farther south.

Temperature in the atmosphere also varies with height. Within the first few hundred meters, temperature significantly changes diurnally. At night, the Earth's surface cools and causes a decrease of the air temperature near the surface. As a result, the so-called *inversion layer* is formed near the Earth's surface. In the inversion layer, temperature increases with height and the *lapse rate* (the rate of decrease of temperature with height) is negative. On the other hand, during the day the Earth's surface is heated by the Sun. The warm

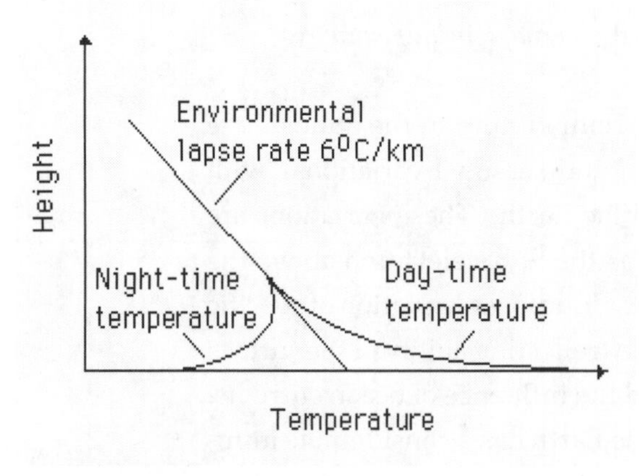

Idealized distribution of temperature near the ground.

surface causes an increase of the air temperature in a thin layer above the ground. As a result, the daytime air temperature near the ground readily decreases with height.

In the troposphere (up to about 10 km above the ground in mid latitudes), the mean temperature decreases with height with a *lapse rate* of about 0.6°C per 100 m (a positive value indicates that the temperature decreases). In the stratosphere, as shown in the figure below, temperature generally increases with height due to an absorption of solar radiation by ozone. In the mesosphere, temperature decreases with height. In the thermosphere, temperature of air molecules increases with height. There is an extremely small number of molecules in this region. Because of low air density, a liquid-in-glass thermometer cannot be used to measure the temperature in the upper atmosphere. Departures from the figure occur due to seasonal and latitudinal variations.

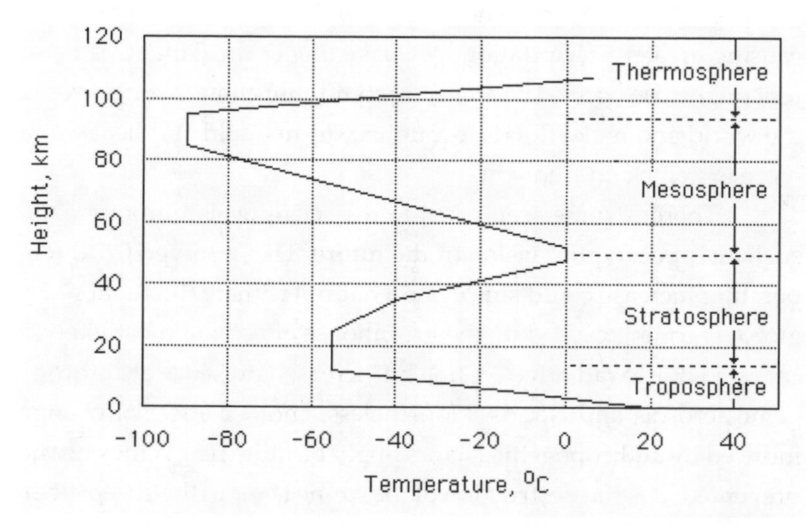

Vertical variation of temperature in the atmosphere

Temporal variations of air temperature can be related to various natural and anthropogenic factors, including volcanic eruptions and increased levels of air pollutants. During volcanic eruptions, tons of dust and ash are spewed into the atmosphere. Some of this material reaches to the levels above the troposphere and is redistributed around the Earth. The temporary presence of volcanic debris causes some of the Sun's energy to be reflected back into outer space before it reaches the Earth's surface. Volcanic particles also intensify cloud formation. More clouds reflect more solar energy back into outer space. Even though the volcanic dust and clouds also prevent some of the Earth's heat from escaping, the resulting effect is a cooling of the Earth. The June 15, 1991 eruption of Mount Pinatubo in the Philippines is believed to have lowered global air temperatures at the Earth's surface by 0.5°C on average during the following year.

On the other hand, there has been a concern (e.g., expressed in the 1995 Report by the Inter-governmental Panel on Climate Change Working Group) that further industrial emissions of CO_2, methane, nitrous oxide and chlorofluorocarbons will cause a *global warming* of our planet (an enhanced greenhouse effect). A 0.5°C surface temperature increase has already been observed since 1880. As a result of the warming, glacier and continental ice could melt, resulting in rising sea levels. More water vapor could be released into the air causing greater precipitation. A weakening of the Gulf Stream current might also occur. The warming could shift climate zones around the world and make floods, droughts, storms, cold and heat waves more extreme and frequent.

There is some scepticism in the scientific community about such an "apocalyptic" vision of the future. The observed 0.5°C temperature increase could simply be a natural climate fluctuation. The global warming caused by the greenhouse effect will most likely be canceled due to radiative cooling by increased presence of anthropogenic aerosols and clouds. Nonetheless, another effect of changes induced by anthropogenic factors might be quite real, if they are not prevented. It is the destruction of the ozone layer in the stratosphere by molecules of chlorofluorocarbons (CFCs). When released into the atmosphere, CFCs slowly move upwards, then break up as a result of solar radiation, produce chlorine molecules which finally react with ozone. Evidence of massive ozone layer destruction, called the *ozone hole*, has been observed over Antarctica. Reduction of the ozone content, and consequently in of the amount of the absorbed solar energy, is expected to cool the stratosphere.

People trust their eyes more than their ears. (Heraclitus)

CHAPTER FIVE

LIGHT

Spectacles in the Sky, Development of Optics, Reflection, Refraction, Law of Refraction, Refraction in the Atmosphere, Mirage, Halo, Camera Obscura, Nature of Light, Speed of Light, Diffraction and Interference, Corona, Polarization of Light, Color, Solar Radiation, Mixing of Colors, Blue Sky, Rainbows, Green Flash, Glories.

SPECTACLES IN THE SKY. Our atmosphere is the ever-changing scene of spectacular optical phenomena. Observation of these phenomena prompts us to investigate the nature of light, the wonderful cause from which they are derived. What is light? This question has no doubt occupied the attention of people throughout the ages. People have looked upward at the sky and have wondered about the meanings of all the things they could see. They observed that when the Sun was rising, the dark sky near the horizon was becoming lighter, then turned red, yellow, and finally blue. When the Sun was setting, the blue sky near the horizon might become green, yellow, then orange, and finally red. At night, the Moon was sometimes surrounded by colored rings of various intensities. On some days, the sky might be enlightened by spectacular tints and rainbows.

DEVELOPMENT OF OPTICS. The word *optics* comes from a Greek word *optikos* which means "visible". *Optics* is the science of studying light. The history of optics began in Greece, two and a half thousand years ago. The ancient Greeks were fascinated with the secrets of light and vision. They studied colors and color phenomena, such as rainbows. They even formulated laws of propagation and the reflection of light.

The Greeks believed that there must be some form of contact between an object of vision and the visual organ. They generally recognized three forms of such contact. Pythagoras (about 600 B.C.) and his followers believed that the human eye was a source of light rays, which propagated on straight lines from an eye to the observed objects. This explanation was generally accepted throughout ancient times. For instance, both Euclid (about 300 B.C.), and Claudius Ptolemy (about 250 A.D.) argued that the eye could send forth a ray or power to the object.

Aristotle (384-322 B.C.) claimed that contact between an eye and an object could be established through a medium called *ether*. According to him, motionless *ether* was black. Any object which could be seen, made the *ether* move. Fast motion produced the bright light. Any other color was a mixture of black and bright. Aristotle maintained that a person could see a green object because the object colored the observer's eye green and this acquisition of color constituted the act of seeing.

Epicures (341-270 B.C.) proposed yet another explanation. According to him, objects could be seen because they could send their images as visual rays through space to the eye.

Theories developed by the Greeks were generally accepted for the next ten centuries. During this time the science of optics made little progress.

From the ninth to the eleventh century A.D., Islamic culture flourished and spread from the Middle East westward to Morocco and Spain. Many of the Greek achievements in optics were translated into Arabic during that period. On Islamic soil, Greek optical tradition was not only reproduced but also enhanced .

About 1025, a great Muslim scholar, known in the West as Alhazen, wrote a famous book on optics in which he corrected the standing theory of eye rays used in the classical period. He proposed that luminous bodies emitted light rays, which entered the eye and stimulated vision. He showed that both incident and reflected rays were in the same plane. Alhazan observed that colors on the surface of a spinning top were perceived as one color, compounded from all of these colors. As mentioned in Chapter 2, he also evaluated the height of the atmosphere based on his observations of twilight.

A study of perspective

The Islamic Empire was of a relatively short duration and political power and cultural influence drifted northward to Europe. In the Middle Ages, the science of optics was called *perspectiva* and was concerned with such matters as the nature and propagation of light and color, the eye and vision, image formation by reflection and refraction, and meteorological phenomena involving light. Most scientists were monks and priests. One of them was the Polish priest, Witelo (ca. 1230-1290), the author of an encyclopedic work on optics *Perspectiva* (ca. 1270).

REFLECTION. In the third century B.C., Euclid first formulated laws of propagation and *reflection* of light. According to his teachings, light was directed in straight lines. Light could be *reflected* , i.e. it could bounce like a ball from surfaces, in such a way that the angle of incidence (between the ray and the normal (perpendicular) line to the reflecting surface) and the angle of reflection remained equal.

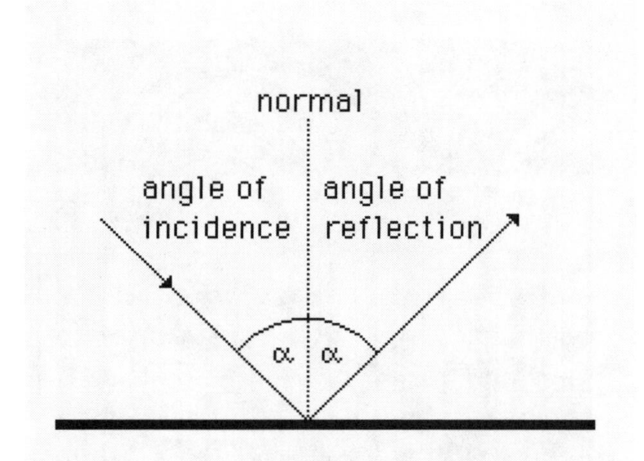

Law of reflection

Three hundred years later, about 50 A.D., Heron, who lived in Alexandria, wrote a book entitled *Catoptrics*. The book contained the theory of mirrors. He believed that our vision was due to some rays emitted from the eye and reflected back by the object. The rays proceeding from the eyes were reflected by mirrors; and the reflections were at equal angles.

5.1. REFLECTION OF LIGHT (1)

Materials: comb, mirror.

Procedure: Hold a comb in the direct line of light. Light should be able to pass through the teeth of the comb. Place a mirror in the path of these rays. Observe that the beams which strike the mirror are reflected at the same angle.

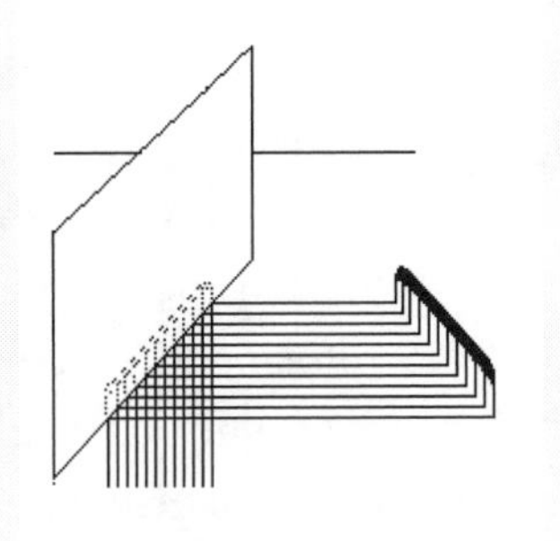

The first application of optics. According to legend, Archimedes of Syracuse in Sicily (287-212 B.C.) was the first to make practical use of the knowledge of optics. During the Roman conquest of Sicily in 214 B.C., he used mirrors to focus the Sun's rays on the boats of invaders, causing them to ignite. The construction was probably similar to that of modern Sun reflectors and consisted of a number of small polished metal mirrors. Presumably, many soldiers simultaneously reflected the Sun's light onto each ship, causing it to catch fire. Contemporary experiments performed in Greece in 1973, by a Greek engineer, who used 70 mirrors, 1.5 x 1 m each, proved that putting a wooden vessel on fire, from a distance of 50 m, was indeed possible.
After the capture of Syracuse, during the second Punic (meaning "Phoenician") war, Archimedes was killed by a Roman soldier who found him drawing a mathematical diagram in the sand. Archimedes was so absorbed in his calculation that he offended the soldier for stepping on his diagram. Marcellus, the leader of the victorious troops, deplored this tragic blunder. Throughout his triumph in Rome, he lamented the death of Archimedes. He also ordered the erection of a splendid monument over Archimedes' grave.

5.2. REFLECTION OF LIGHT (2)

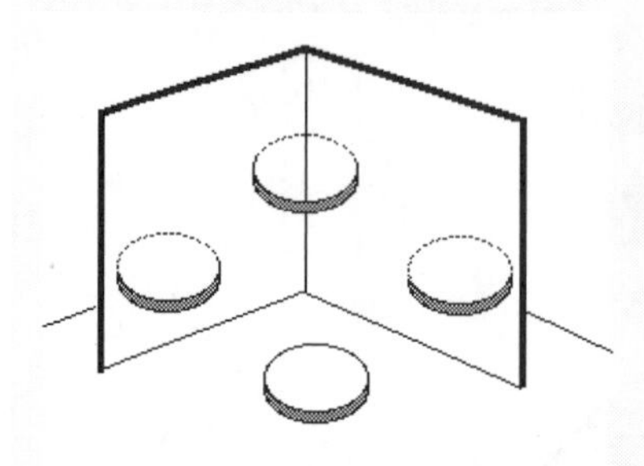

Materials: two mirrors, tape.

Procedure: Tape two mirrors together. Place any object between the mirrors. Multiple images will be formed in the mirrors. Set up the mirrors so that they are at various angles to each other. Observe that the number of images depends on the angle of the mirrors.

REFRACTION. In the second century B.C., Claudius Ptolemy, who lived in the temple of Serapis in a suburb of Alexandria in Egypt, investigated light as it passed from one transparent substance to another. He described these observations in his treatise *Optica*. The original manuscript was lost, but the Arabic translation survived. In the twelfth century, the Arabic version was translated into Latin.

According to Ptolemy's observations, visual rays were altered not only by *reflection*, i.e. rebounding from objects, by also by *refraction*, i.e. bending by transparent media which permitted penetration. Ptolemy illustrated the phenomenon of refraction by placing a coin at the bottom of a vessel filled with water. In his studies, Ptolemy used an apparatus, which is employed to this day, to study refraction. The instrument consisted of a vessel, a scale to measure angles, and three pointers I, R, and N (see the figure).

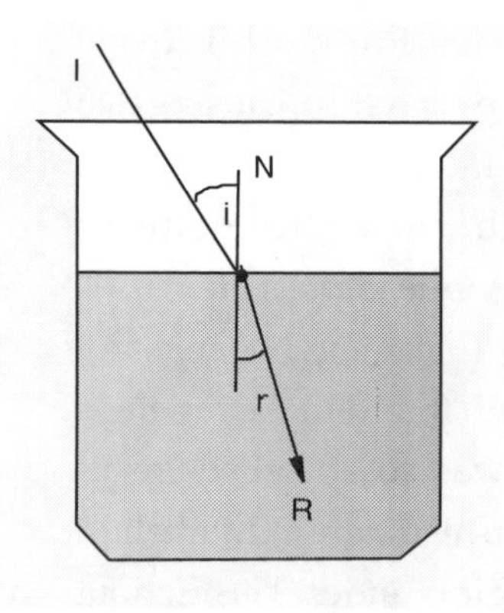

Refraction

From a series of experiments and measurements of incident angle i, and refraction angle r, Ptolemy obtained the following results:

i (air)	r (water)	i (air)	r (water)
10°	8.0°	50°	35.0°
20°	15.5°	60°	40.5°
30°	22.5°	70°	45.5°
40°	29.0°	80°	50.0°

Ptolemy concluded that the refraction angle r in water was smaller than that in air. However, Ptolemy was not able to discover the exact relation between angles i and r. He only proposed an approximate relationship: $r = 0.825i - 0.0025i^2$. The exact solution to the problem remained unsolved for the next fifteen centuries.

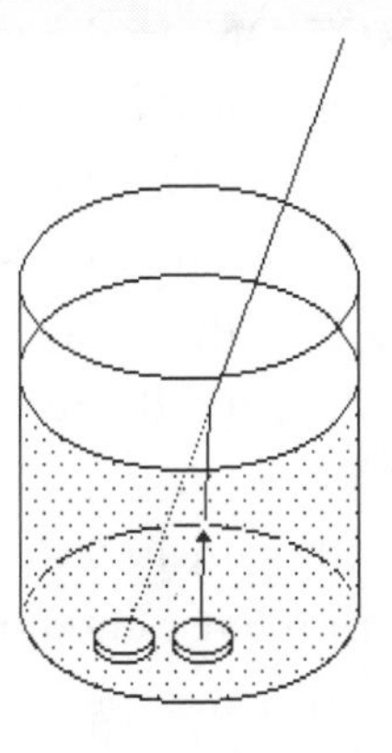

5.3. REFRACTION OF LIGHT (1)

Materials: cup, coin, water.

Procedure: Place a coin at the bottom of the cup, near its edge. Look at the cup from above so that you cannot see the coin. Pour water into the cup carefully so that the coin remains in position. At some point, the coin which was previously invisible will be seen. The same eye position relative to the cup must be maintained throughout the experiment.

5.4. REFRACTION APPARATUS

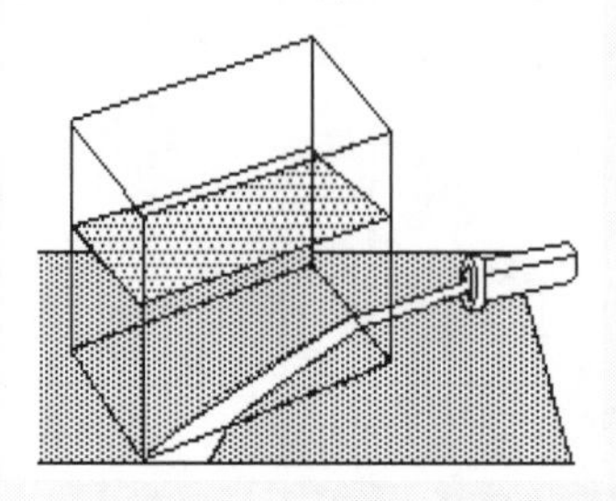

Materials: aquarium, water, drop of milk, flash-light, protractor.

Procedure: Fill the aquarium with water. Add a drop of milk in the water for better observation. Shine a beam of light through the glass wall of the aquarium in a darkened room. Measure the angles of incidence and refraction with a protractor. Shine a beam which passes from air to water and to air again. Notice that there is the *critical angle* α_c, at which light follows the interface between water and air. At angles greater than α_c, light undergoes internal reflection and cannot refract out of water.

LAW OF REFRACTION. In the year 1620, Willebrord Snell (1581-1626), a Dutch professor of physics at the University of Leyden (and a teacher of Otto von Guericke), formulated the law of refraction. According to this law, for any given pair of transparent media, the ratio $\mathbf{n} = \sin(\mathbf{i}) / \sin(\mathbf{r})$ is constant, where **i** and **r** are the incident and the refraction angles, respectively. The constant **n** is called the *index of refraction*.

Snell correctly concluded that light traveling from a less-dense to a more-dense medium loses speed and bends toward the *normal* (the line perpendicular to the border between both media), while light entering a less-dense medium increases in speed and bends away from the normal. Snell's manuscript was published in 1662, after the author's death.

Refraction

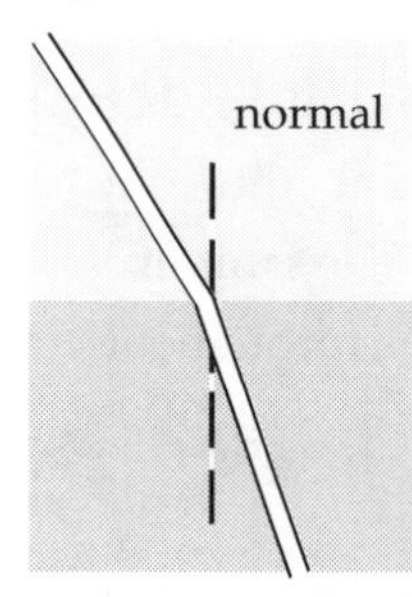

The same law was derived differently by René Descartes in his *Dioptrique* (1637). Descartes argued that light moved like a ball. Nevertheless, he incorrectly assumed that light entering the water increased its velocity from **U**, to **V**, while its tangent component **W** was unchanged (see the figure below). From this Descartes obtained that $\mathbf{W}/\mathbf{U} = \sin(\mathbf{i})$, $\mathbf{W}/\mathbf{V} = \sin(\mathbf{r})$. After elimination of **W**, this yielded the same form of the refraction law: $\sin(\mathbf{i}) / \sin(\mathbf{r}) = \mathbf{V}/\mathbf{U} < 1$. Even though his result formally agreed with Snell's law, it was incorrect because, as it was proved later, the speed of light in air is larger than its speed in water.

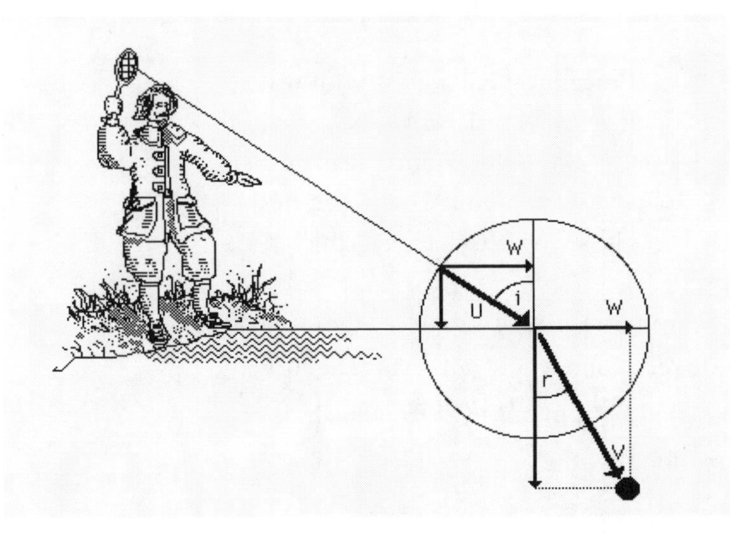

Descartes' incorrect explanation

Using Ptolemy's measurements on page 109, and Snell's law, the index of refraction **n** for water can be estimated as:

i	r	n	i	r	n
10°	8 °	1.25	50°	35°	1.34
20°	15.5°	1.28	60°	40.5°	1.33
30°	22.5°	1.30	70 °	45.5°	1.32
40°	29°	1.32	80°	50°	1.29

The average value of his results was **n** = 1.304. The modern value is **n** = 1.333. The accuracy of Ptolemy's measurements was 0.5 of a degree and explains the variation in the values of **n** in the table.

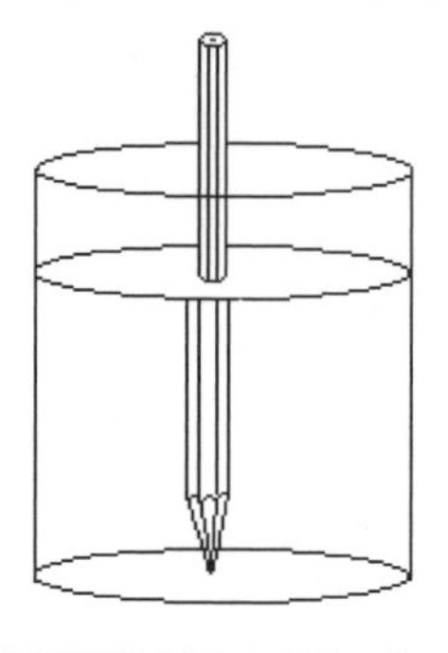

5.5. REFRACTION OF LIGHT (2)

Materials: glass, pencil, water.

Procedure: Fill a glass with water. Insert a pencil. Move the pencil in the glass closer and then farther away from you. The pencil in the glass seems to change its thickness.

Explanation: If the glass surface were not curved the effect of magnification due to refraction would not occur. The curved surface directs refracted light to our eyes from such directions that an impression that the pencil is thicker appears.

REFRACTION IN THE ATMOSPHERE. Changes of the refraction index, caused by fluctuations of temperature and humidity in the atmosphere, are responsible for the twinkling of stars. Refraction in the atmosphere also affects the position of celestial bodies in the sky. The celestial bodies appear to be slightly higher above the horizon than they really are.

The following table can be used to explain the flattening of the Sun and the Moon near the horizon (also see the figure on the next page). The table indicates that the bending angle is largest near the horizon:

Apparent elevation angle	Deviation from true	Apparent elevation angle	Deviation from true
0.0°	33.8'	40°	1.1'
0.5°	28.2'	50°	47.9"
10°	5.2'	60°	33.0"
20°	2.6'	70°	20.1"
30°	1.6'	80°	10.1"

Both, the Sun and Moon have angular diameters of about 0.5°. As seen in the table, at the horizon, the lower edge of the Sun's disc appears 33.8' higher than it actually is, while the top edge is only 28.2' higher. As a result, the disc appears elliptical with the vertical axis being 33.8'- 28.2'= 5.6' shorter than the horizontal one. The flattening amounts, therefore, to about 5.6'/30' = 20% of the Sun's diameter. Since the Sun moves about 180° per 12 hours = 15' of arc per minute, an observer at the equator would see the Sun about two minutes (30'/15' = 2) before it actually rises, and two minutes after it actually sets behind the horizon.

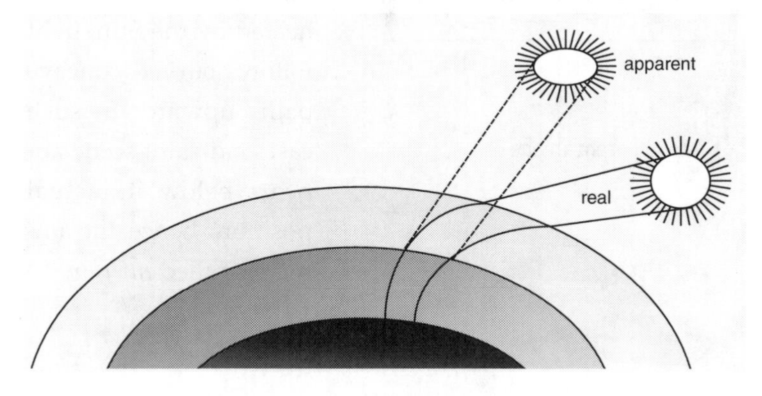

Flattening of the Sun due to refraction

Vision is not merely a static copy of the world. Our eyes are not ideal visual instruments. Sometimes, we see the world

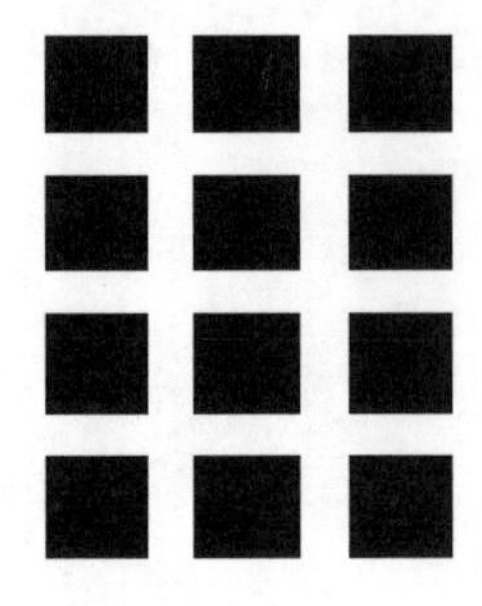

Example of an optical illusion

differently from what it really is and call it *illusion*. Notice, for example, in the accompanying figure, that the spaces between black squares appear darker. That is illusion. Voltaire (1694-1778), the French writer and philosopher, thought that "*Illusion is the first of all pleasures*". One of the most interesting natural illusions is the apparent size of a full Moon when seen near the horizon. A few hours later, when overhead, it seems much smaller. The phenomenon was already studied by Carl Friedrich Gauss (1777-1855), the eminent German mathematician. Gauss correctly concluded that the Moon near the horizon only looks greater in comparison with terrestrial objects, such as trees and houses. Photographs confirm that the diameter of the Moon's disk is the same in both cases.

MIRAGE. Mirages are distorted optical images of the sky and the Earth's surface. Mirages appear over deserts, oceans or snow fields. They are caused by refraction and come in a wide variety of forms.

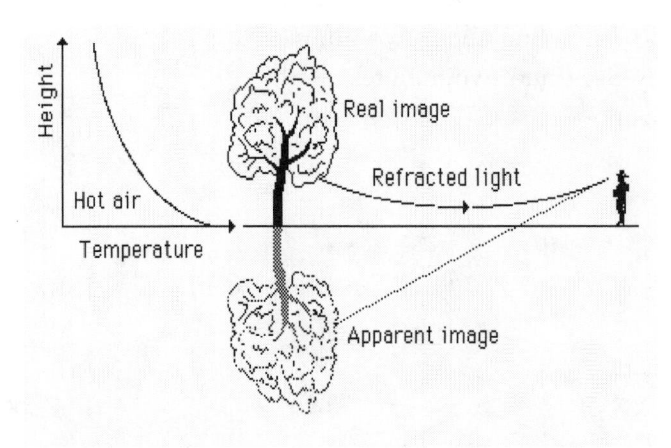

Inferior mirage

When the air temperature decreases with height (the thinner air is near the ground), as during the day over surfaces heated by the Sun, light follows curved concave paths upward. In such cases a distant scene appears below its actual position; hence the mirage is called *inferior*.

Discovery. There are suggestions that the Celts discovered Iceland by seeing its mirage from Faroe Islands located 385 km away from Iceland. Greenland, which is located 300 km away from Iceland, could have been discovered in the same way.

When the air temperature increases with height (the denser air is near the ground), as when warm air flows over very cold air surfaces, light rays follow curved concave paths downward. In such cases a distant scene appears above its actual position; hence the mirage is called *superior*.

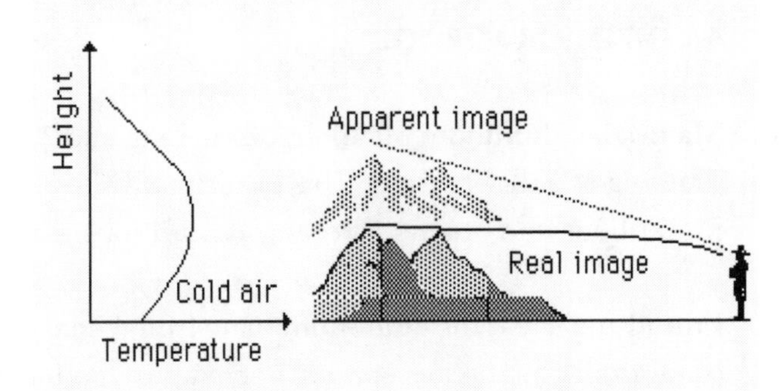

Superior mirage

A special kind of a superior mirage is the *Fata Morgana*, first recorded as seen across the Straits of Messina, between Italy and Sicily. Fata Morgana transforms a horizon into a vertical wall with columns and spires. In Italian, *Fata Morgana* means "fairy Morgan". According to legend, Morgan was the half-sister of King Arthur, who lived in a crystal palace beneath the water and had the magical powers to build castles out of thin air.

Nature's practical jokes. In 1818, two English sailors, brothers James and John Ross, reported that they had seen from their ship an unknown land covered by a high mountain range near Baffin Island in Northern Canada. About twenty-five years later, the same unknown land was spotted by Admiral Matthew C. Perry, who named it "Crocker Land". In 1913, the American Museum of Natural History commissioned Donald B. MacMilan to lead an expedition to explore this unknown frontier. At first, the result of MacMilan's expedition was disappointing. Crocker Land had disappeared. Only open water was found where it was supposed to have been. MacMilan sailed around the designated spot. Finally, about 200 miles west, he saw a land covered by a huge mountain range. MacMilan sent a small crew of men ashore. As the team moved toward the mountains, to their astonishment, the range of mountains moved away. The faster they walked, the faster the mountains receded. When they stopped, the mountains did not move either. The chase continued throughout the entire day, and finally, after the Sun had set, the range disappeared. The men looked at each other, and suddenly realized that they had become the victims of one of nature's practical jokes. Crocker Land was a *mirage*.

5.6. INFERIOR MIRAGE

Materials: aluminum plate, about 3 m long and 20 cm wide, gas torches, sand, drawing of a desert scene. This experiment should be performed with special precautions and only in laboratory conditions.

Procedure: Heat the aluminum plate from beneath by gas flames. The flames come from gas torches spaced 10 cm apart, connected to a gas line. A desert scene can be placed at one end of the heated plate and viewed as an inferior mirage from other end. To ensure there is no reflection from the plate, the plate should be covered with sand.

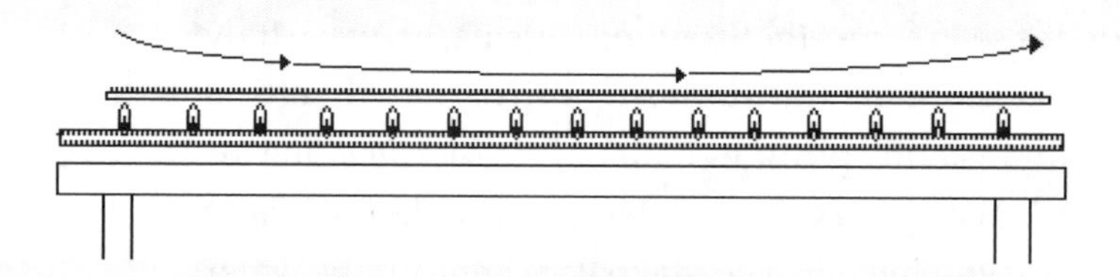

Comment: Mirages can also be seen on hot summer days along walls with a length of 10 m or longer, which are facing the Sun. If you place a bright object at one end of the wall, a few centimeters from its surface, and look at it from the other end of the wall, you will see the object reflected as it would from a mirror.

5.7. SUPERIOR MIRAGE

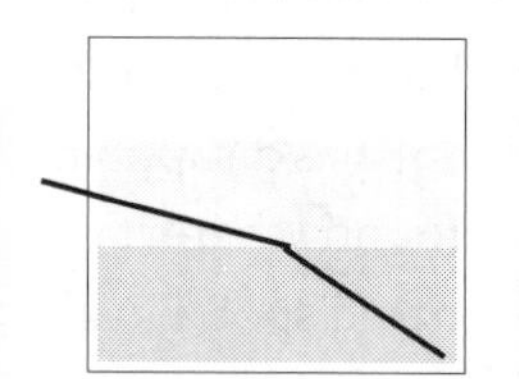

Materials: tank constructed with sides of glass and ends of clear plastic 60 cm long, 30 cm high, and 8 cm wide, water, salt, flashlight.

Procedure: Fill a tank about half full with saturated salt water. Introduce a layer of fresh water on top, by slowly pouring it on a floating board. As a result, the fresh water does not mix with the salt layer. Shine light through the tank. Observe light bending.

Explanation: The experiment demonstrates a mechanism of a superior mirages. During a superior mirage, rays passing through air layers move in curved concave paths downward.

HALO. The Sun or the Moon is often encircled by spectacular rings of light called *halos*. The phenomenon is caused by the refraction through ice crystals suspended in clouds, about 10 km above the Earth's surface. The most common is the 22° halo, a ring of light 22° around the Sun or the Moon. Less common is the 46° halo, a ring of light 46° encompassing the Sun or the Moon.

The shape of a halo is determined by shapes of ice crystals and their orientation in space. The 22° halo is due to refraction by falling ice crystals which are randomly oriented in a plane perpendicular to a ray of incident sunlight. The 46° halo is due to refraction by falling ice crystals with hexagonal column shapes which are oriented as in the figure above. Haloes can have coloration (with red inner and blue outer rings) because of the dependency of refraction upon wavelength.

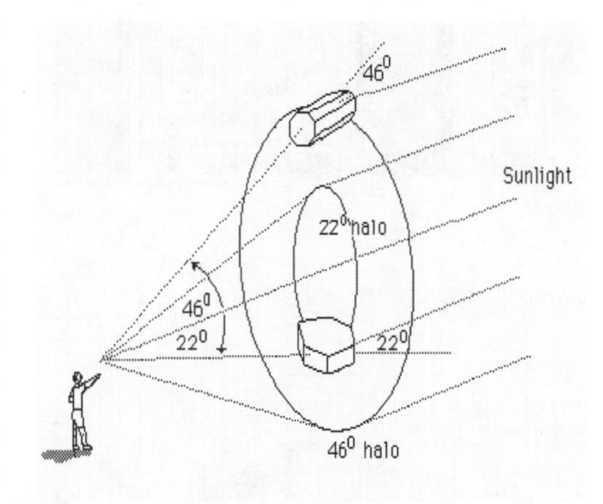

22° and 46° halo images

There is a large variety of images during a halo, called *arcs, circles* and *sundogs. Sundogs* are colored spots on each side of the Sun. Some of the possible halo arcs, circles and sundogs around the Sun are shown in the figure.

A halo can be seen at any time of the year. The practical value of halos for weather forecasting is due to the fact that halos appear on high level clouds which often precede *warm fronts,* and may indicate potential precipitation within 24 - 48 hours. An ancient rhyme predicts: *halo on the sun or moon, snow or rain are coming soon.*

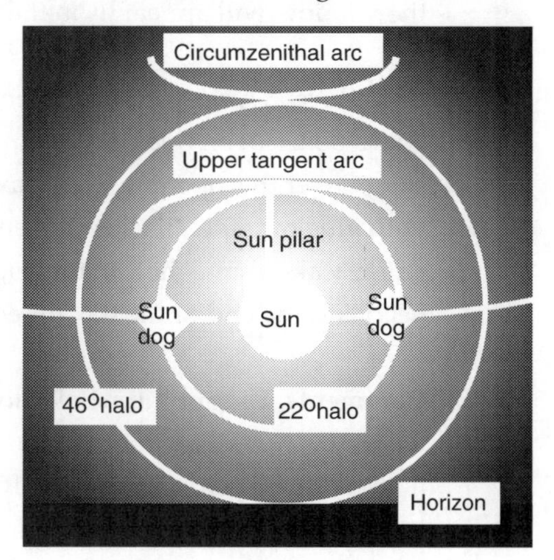

Variety of halo images

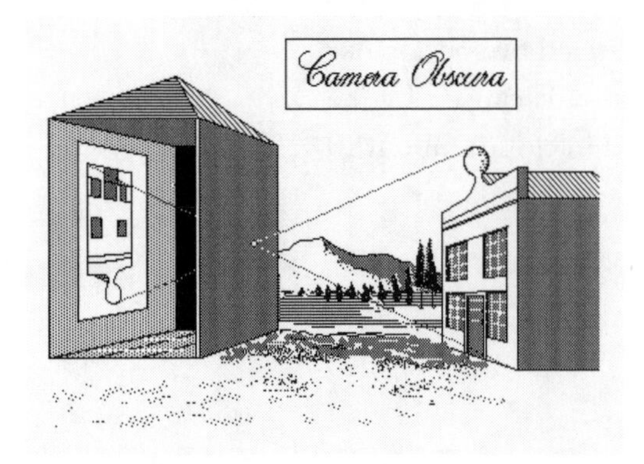

Camera obscura

CAMERA OBSCURA. In order to understand vision, Leonardo da Vinci (1452-1519) dissected a bull's eye. He also constructed a model of the eye, which he called *camera obscura* (the other name is *pin-hole camera*). *Camera obscura* in Italian means a "darkened chamber". It was so named because the first pinhole camera was made in a darkened room.

5.8. THE PINHOLE CAMERA

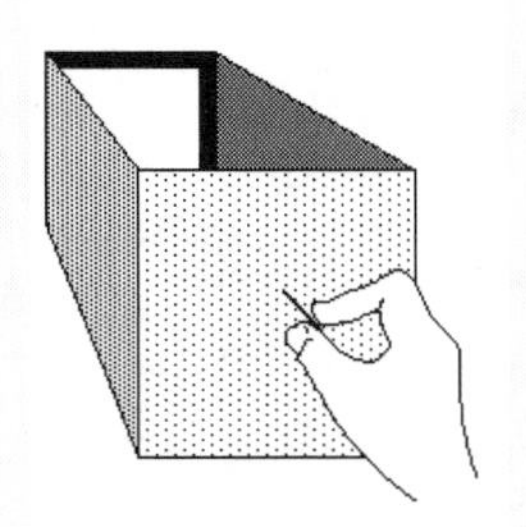

Materials: cardboard, screen.

Procedure 1: Select a room with a small window. Darken the room by placing a cardboard over the window. Make a small hole, about 1 mm in diameter, in the cardboard. Place a white screen, about 1 m away from the window. On the screen, a grayish, up-side-down image of the scenery outside of the window will appear as in the figure above.

Procedure 2: A similar experiment can be performed using a cardboard box. Cut an opening, about 5 cm square, in the center of one side. Cover this opening with waxed paper and attach it to the box with glue or tape. Blacken the entire inside of the box with black paint. Make a small hole in the opposite side of the box. Cover your head and the end of the box to see the images clearly.

Comments: The optimum hole radius is approximately $r = \sqrt{(0.6\lambda d)}$, where λ is the wave-length of the light and d is a distance from the hole to the screen. For $\lambda = 5 \times 10^{-7}$ m (center of the light's visible range) and for d = 1 m, r = 0.5 mm. Larger holes give less resolution, while smaller holes produce diffraction patterns (see page 123).

By looking at the shadows of trees during the day, many bright spots projected on the ground will be seen. These are pinhole images of the Sun. They are made when light passes through the tiny spaces among the leaves. During partial solar eclipses, these sun-fleck images on the ground become crescent-shaped.

NATURE OF LIGHT. The Dutch physicist, Christiaan Huygens (1629-1695), proposed in his treatise *Traité de la Lumiére* (1678) that light traveled as *waves*. Huygens argued that light was transported in the same way circular waves moved away from the place where a pebble hits the surface of the water. Huygens' theory explained reflection, refraction, and several other properties of light. It also predicted that light traveled more slowly in glass or water, than in air.

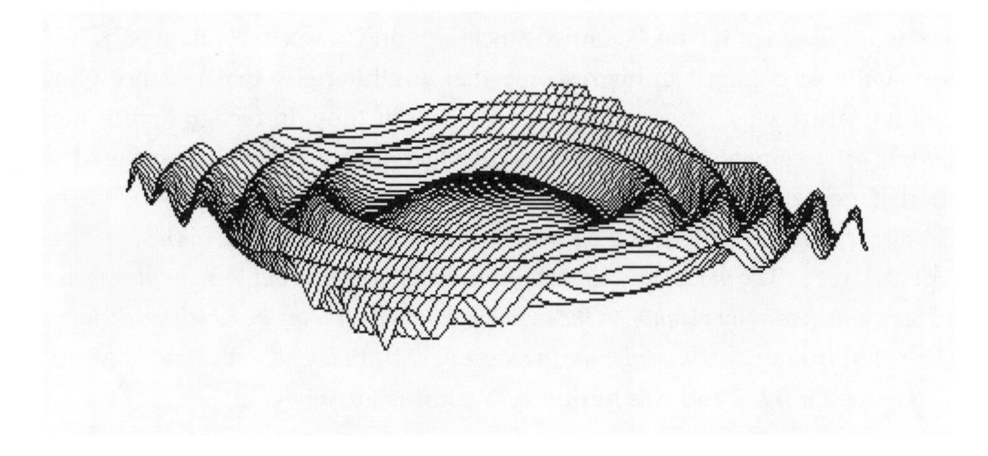

Surface waves

In the late seventeenth century, Isaac Newton (1642-1727) attempted to explain the properties of light by suggesting that a light beam consisted of a stream of tiny particles, which he called *corpuscles*. Corpuscles supposedly traveled in straight lines. Newton thought that the reflection of light by a mirror was due to corpuscles bouncing off the surface of the mirror. He also believed that the corpuscles moved with different velocities in various substances.

Isaac Newton (1642-1727)

Isaac Newton was born prematurely on 25 December 1642, in the village of Woolsthorpe, England. His father died a few months before Isaac's birth. In 1661, on the advice of his uncle, Isaac entered Trinity College in Cambridge. During the plague, between 1665-1666, by Newton's own words, he "*was in the prime of his age for invention of mathematics* (e.g. he invented calculus) *and philosophy more than at any time since*" In 1669, he became a professor at Trinity College. Few students attended his lectures, and even fewer understood him. He often lectured to the walls. Newton was known as a scatterbrain. He ate sparingly and often forgot to eat at all. He was a bachelor, whose manner of dress was unkempt. He appeared with shoes worn down at the heels, untied stockings, and scarcely combed hair. He was usually so engaged in his research that anything else that was not of a scientific nature was viewed simply as a waste of time. In recognition of his scientific achievements, he was knighted by Queen Anne. In 1693, he suffered a period of mental depression. Newton entered Parliament in 1701. From 1703 to 1727, he served as president of the Royal Society. On his deathbed, his last words allegedly were: "*I do not know what I may appear to the world; but to myself I seem to have been only like a boy playing on the seashore, and then finding a smoother pebble or prettier shell than ordinary, whilst the great ocean of truth lay all undiscovered before me*". He died in 1727 and was buried in Westminster Abbey.

Controversy surrounding the rivaling wave and corpuscular theories, raged on for about a century. However, conclusive proof could not be found because of the lack of definitive experimentation. In 1801, Thomas Young (1773-1829) discovered dark and light interference fringes (lines) which could be explained only in terms of the wave theory. Later, in 1850, Jean Bernard Foucault (1819-1868) proved that light traveled slower in water than in air, and that Huygens was correct. However, this did not mean that

Newton was wrong. According to modern physics, light is of a dual nature and simultaneously behaves as a stream of corpuscles (photons) and waves.

SPEED OF LIGHT. For a long time, the speed of light remained a controversial issue. Some believed it was finite while others believed it was infinite. For instance, in 1634, René Descartes wrote: "*My understanding is that light is transferred instantaneously in space and if experience will show contradictory, I am ready to admit that I know absolutely nothing about physics* ". Certainly, he did not understand everything.

In his treatise *Traité de la Lumiére*, Christian Huygens (1629-1695) argued that the speed of light was very fast, but finite. Huygens was familiar with the experiment first performed by the Danish mathematician Olaus Rømer (1644-1710). Rømer observed the eclipses of Io (in which Io hides behind Jupiter), the largest of the four moons of Jupiter, and discovered in 1610 by Galileo. Rømer noticed that the intervals between two consecutive eclipses decreased when the Earth approached Jupiter, and increased when the Earth receded.

Jupiter and Io

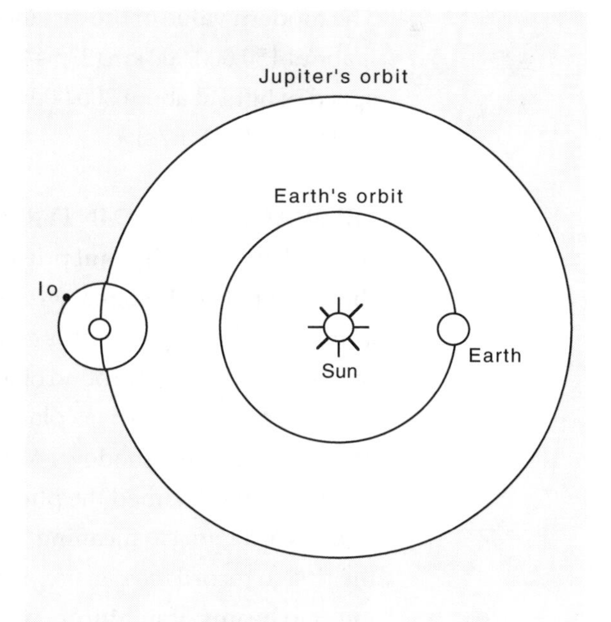

Jupiter encircles the Sun in 12 years, and its position with respect to the Sun does not change much during one Earth year. Io encircles Jupiter in about 42.5 hours, (exactly 1 day, 18 hours, 28 minutes and 35.9 seconds). Unfortunately, this is not a very convenient period for observation. If an eclipse occurs at 7 PM in the evening, the next eclipse will occur in the afternoon, around 1 PM two days later, when Jupiter is not visible in the daytime sky. However, every fourth eclipse can be observed at night every 7 days, 1 hour, 54 minutes and 23.6 seconds, over a period of several weeks.

Eclipses of Io can be treated as light signals sent in equal time intervals. If the speed of light were infinite, no differences would be observed. Rømer observed that the intervals changed as the distance between Jupiter and Earth changed. The greatest change that Rømer was able to measure during ten consecutive years was about 1000 seconds (about 17 minutes). By analyzing his data, he was able to conclude that the difference was caused by the fact that the speed of light was finite and that the time to travel a distance of double the radius of the Earth's orbit around the Sun was 1000 seconds. Rømer did not give the numerical value of the speed of light.

In 1672-1673, three French astronomers, J. D. Cassini, J. Picard and J. Richer, evaluated the diameter of the Earth's orbit as being equal to 21,600 diameters of the Earth. They accomplished this result through simultaneous observations of Mars from two points on the Earth, i.e., Paris (France) and Cayenne (Guyana). The modern value of the distance between the Sun and the Earth is about 150,000,000 km (23,544 Earth's diameters). This yields the speed of light at about 300,000 km/s (the exact value in a vacuum is 299,792,458 km/s).

DIFFRACTION AND INTERFERENCE. In 1665, Francesco Grimaldi (1618-1663), a Jesuit priest and professor of mathematics at the University of Bologna, wrote a Latin treatise on light, *Physico-Mathesis de Lumine, Coloribus et Iride.* In this work, he recognized that light could bend around objects. As an example, he described that when a thin stick was placed in the way of a solar beam in a *camera obscura,* the shadow of the stick was thinner than the stick itself. Grimaldi named the phenomenon *diffraction,* based on the Latin word *diffracto* meaning "breaking up". Grimaldi was also the first to record *interference,* an effect produced by superimposing two beams of light from a pinhole camera with two minute pinholes. *Interference* takes place when waves arrive at the same point in space. When a wave peak overlaps a peak, and a trough overlaps a trough, the ripples are reinforced by "constructive" interference. When peak meets trough, they cancel out by "destructive" interference (see Experiment 5.11).

In 1801, Thomas Young (1773-1829) delivered a lecture to the Royal Society in London on light and color. He indicated the significance of Grimaldi's experiments as a strong support for the wave theory of light (he also introduced the term *interference).*

Many scholars opposed Young's theory because it contradicted the authority of Newton. Young was ironically described as thinking in a "wavy" way (certainly, his regular performances with a circus as a tight-rope walker were not helpful in gaining prestige as a scientist). As a result of the criticism, Young ceased publishing for a period of time. *"Great spirits have always encountered violent opposition from mediocre minds"*, Einstein once said.

5.9. DIFFRACTION FRINGES

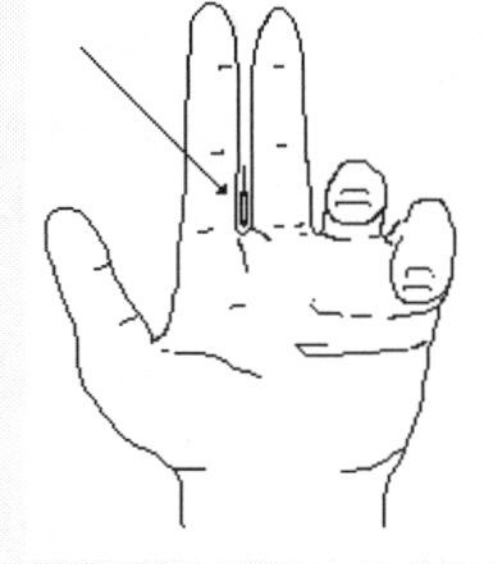

Procedure: Look through a space between the beginning of two fingers which are almost touching each other. Dark lines will be seen. These dark lines are dark fringes, caused by diffraction (bending) and interference (canceling) of light passing through the slit.

Comment: The diffraction patterns can also be observed on CD-ROM disks. The tiny, diffuse spots floating sometimes in your field of vision are diffraction patterns as well. They are formed when light passes spherical blood cells on the vitreous humor, the jellylike tissue that fills the ball of the eye.

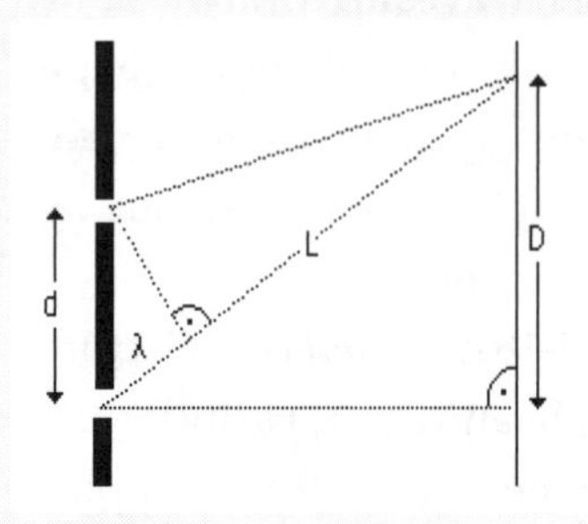

5.10. DIFFRACTION

Materials: negative slide with two parallel lines, light source.

Procedure 1: Make a negative slide by photographing two parallel lines. The two lines on the slide should be as close as d = 0.5 mm apart or even closer. After the slide is ready, place it a few meters away from a light source with a straight filament (for example, a car light bulb). The two slits should be parallel to the filament. On the translucent screen (in order to see better), Young's diffraction fringes can be observed a few meters beyond the slide.

Procedure 2: Select a piece of paper. Cut two parallel slits in it by using two razor blades with a thick paper in between. By looking at light through both slits, alternating light and dark lines can be seen.

Comment: Measure the spacing of slits d, the bright bands D, and a distance from the slits to the fringes L. The wavelength of light λ can be estimated using the geometry sketched in the figure: $\lambda / d \approx D / L$, or $\lambda = d D/L$.

5.11. INTERFERENCE

Materials: two prepared transparencies obtained by photocopying the circle patterns shown on the next page, overhead projector.

Procedure: Prepare two transparencies. Superimpose both transparencies on an overhead projector. The two-source interference patterns will be produced. Each pattern represents a wave from a point source. Spacing between adjacent circles is equal to the wavelength.

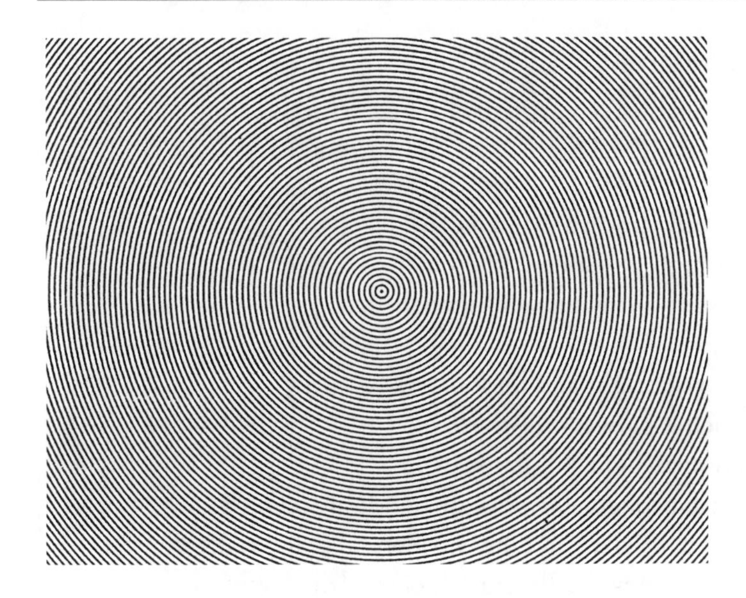

Circle patterns used in Experiment 5.11

CORONA. When the Moon is seen through a thin veil of clouds composed of tiny water droplets, a bright ring of light, called a *corona*, may encircle it. The word *corona* comes from the Latin word meaning "crown". The same effect can occur around the Sun, but it is more difficult to notice because of the Sun's brightness.

The phenomenon may appear as white rings, or on some occasions as colored rings (with blue rings on the inside and red ones on the outside). The rings do not extend farther than about four to twenty moon diameters (2° to 10°) with two to three rings. Coronas appear on mid or lower level clouds which precede warm fronts. Therefore, they may indicate potential precipitation within 12-24 hours.

A corona around the Moon

A corona is caused by the diffraction of light. Where light waves constructively interfere, a bright light is seen, otherwise, it is black in color. Diffraction causes bending of light waves as they move along boundaries of objects (e.g., water droplets).

5.12. DIFFRACTION BANDS

Materials: feather.

Procedure: Look carefully towards (but not directly at) the Sun through a feather. The color diffraction bands caused by diffraction will appear.

Comment: The position of bright and dark bands depends upon the spacing between feather's elements and the wavelength of light. Similarly, in the case of a corona in the atmosphere, the position of bright and dark rings depends upon the drop sizes and the wavelength of light. If droplets little differ in size, rings of different colors are formed with red rings on the outside and blue rings on the inside.

POLARIZATION OF LIGHT. In 1669, Erasmus Bartholinius (1625-1692) of Denmark reported that when objects are viewed through a calcite crystal, called Iceland spar, they appear double. The phenomenon was named *polarization* by the French physicist, Étienne-Louis Malus (1775-1812), who wrongly assumed that light might have different poles as magnets do. In 1817, the phenomenon was explained by Augustin-Jean Fresnel (1788-1827), who suggested that light waves oscillate in various directions at right angles to the beam's path (as shown in the figure). If the vibrations are confined to just one plane, the light is polarized. Calcite divides the beam into two polarized beams which are refracted differently. Polarization proves that light performs a transverse rather than a longitudinal wave motion, because longitudinal oscillations along the beam's path can not be polarized.

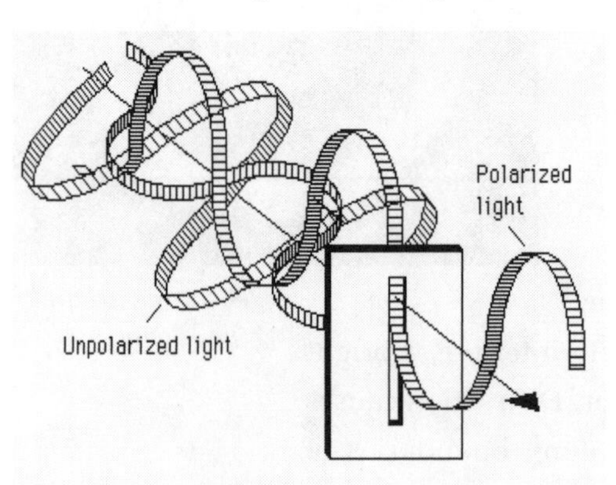

Polarization of light

Polarized light. Polarized light occurs in nature to large extend. For example, light reflected from smooth surfaces is polarized, as well as the light from a rainbow. The light from the blue sky is polarized, especially from a direction at right angles to the direction from the observer to the Sun. These effects can be observed with the help of a photographic polarized filter, or polarized sunglasses. Honeybees, ants and many animals use the polarization of light as a navigation instrument. Photoreceptors in insects' eyes are able to detect the polarization of ultraviolet light in the atmosphere. As a result, insects can orient themselves with respect to the sky.

5.13. POLARIZATION (1)

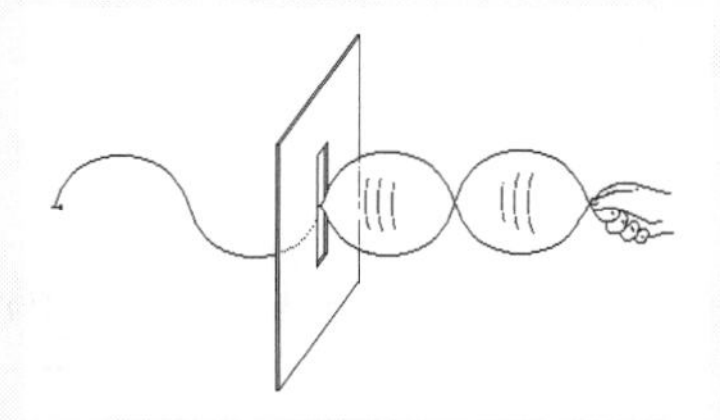

Materials: rope, piece of wood with a vertical slit.

Procedure: Tighten the rope to the wall. Pass the other end of the rope through the slit. Holding the free end of the rope, force it into circular motion. Notice that the vertical slit transmits only transverse waves in the vertical plane. Thus, it acts as a vertical polarizer for waves along the rope.

5.14. POLARIZATION (2)

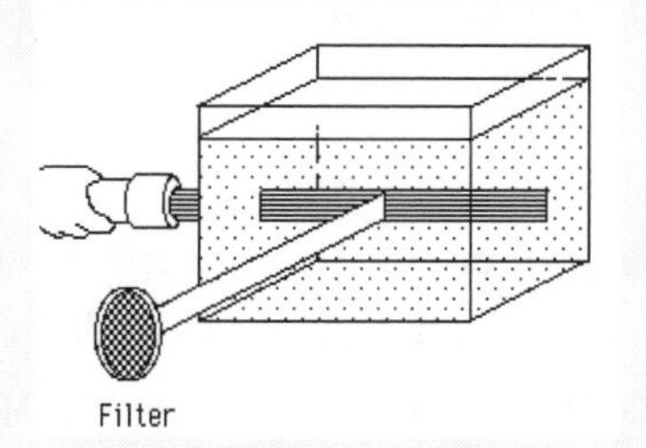

Materials: plastic storage box, milk, flash-light or overhead projector, photo-camera polarization filter.

Procedure: Fill the plastic storage box with water. Add a few drops of milk, so that the beam of light passing through the water becomes clearly visible. For the best visibility, the light source should be located near the side of the box rather than near the middle. Rotate the polarization filter. The light disappears (twice on one complete rotation) for some orientations of the filter, and reappears (also twice) at others, showing that the scattered light is polarized.

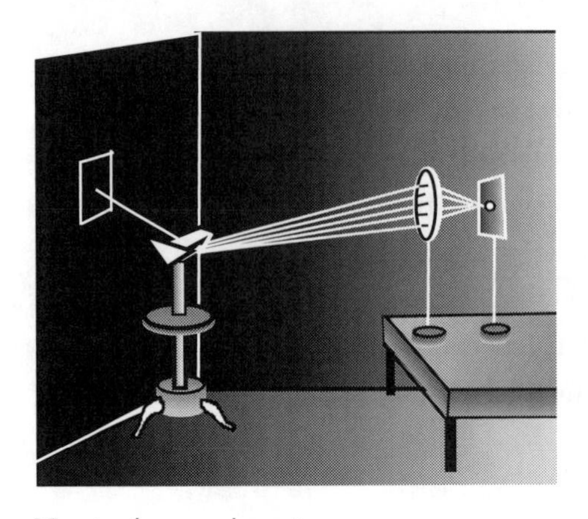

Newton's experiment

COLOR. In 1666, Isaac Newton experimented with a pinhole camera and glass prisms. He allowed a narrow beam of sunlight, about 1 cm wide, to enter the room through a small hole in a window shade. Then he placed a triangular glass prism in the light beam. As a result, he noted that the prism split the beam into a number of different colors: red, orange, yellow, green, blue and violet, as in the figure below. He called this band of light a *spectrum*.

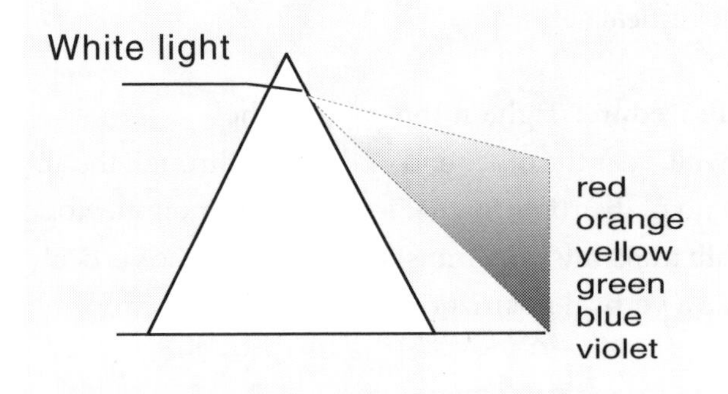

Prism refraction

A scientific argument. In 1672, young Isaac Newton submitted a paper on color to the Royal Society. The paper was reviewed by Robert Hooke, a supporter of the wave theory, who rejected it. Newton was quite hurt. After Newton published his *Principia* in 1687, Hooke allegedly complained about not receiving more credit for his contributions in optics. Hooke denied these rumors, and in response wrote a letter to Newton implying that Henry Oldenburg, the secretary of the Royal Society, was responsible for spreading false reports. Newton replied with a conciliatory letter, which included these words: "*What Descartes did was a good step. You have added much in several ways. If I have been able to see a little further than some others, it was because I stood on the shoulders of giants.*"

Newton proved that the effect of color splitting was caused by refraction, and that each color in a spectrum had its own angle of refraction. Newton also demonstrated that divided colors could be recaptured by the second prism, which merged them back into a white light. He found that the second prism did not produce any further dispersion of the "homogeneal" (monochromatic) light. As a result, he concluded that "*light itself is a homogeneal mixture of differently refractable rays*". Newton thought that each color of light was made by different corpuscles.

5.15. SPECTRUM

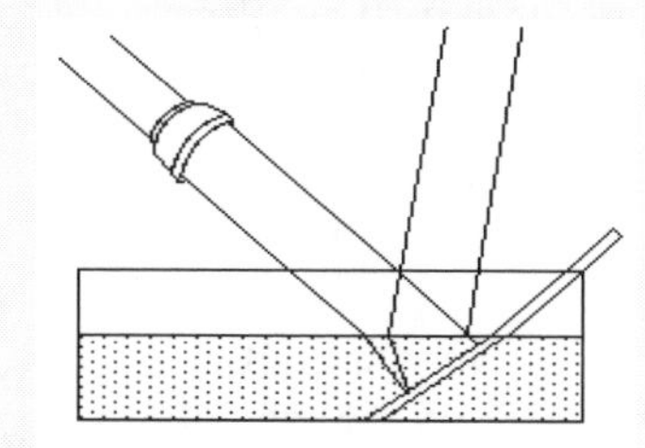

Materials: bowl of water, small mirror, flashlight.

Procedure: Place a mirror in a bowl of water, about 30° to the surface of the water. Darken the room and then shine a flashlight toward the mirror. A small band of colors will appear on the ceiling.

Explanation: The water acts as a prism, refracting each color at a slightly different angle.

SOLAR RADIATION. In 1777, Carl Scheele (1742-1786) investigated silver chloride, known to eighteenth century scientists as a substance which changed in color from white to purple when exposed to light. Scheele found that the color change occurred more quickly toward the violet end of the spectrum. Based on this observation, in 1801, Johann Wilhelm Ritter (1776-1810) discovered the invisible rays beyond the violet end of the spectrum. These invisible rays were named *ultraviolet*.

Ultraviolet radiation. Ultraviolet radiation enables our bodies to synthesize vitamin D, required for strong bones and teeth. The lack of it results in severe bone deformities. Most foods provide little vitamin D. However, our bodies can produce vitamin D when exposed to sunlight. On the other hand, too much vitamin D leads to brittle bones and calcium deposits which could result in kidney stones. Also, overexposure to ultraviolet radiation can cause cancer. Therefore, our bodies have self-defense mechanisms regulating the amount of ultraviolet light which can enter skin cells. This protection is accomplished by the color of our skin. People living in areas where amounts of sunlight are high, have darker skin. The dark skin blocks out most of the ultraviolet radiation. On the other hand, people living in areas where amounts of sunlight are low, have lighter skin. The ultraviolet radiation can easily penetrate light skin.

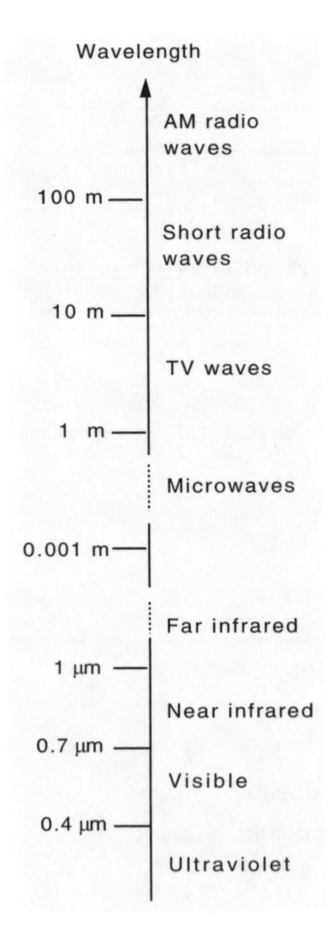

In 1800, William Herschel (1738-1822) carried out various experiments concerning the heating power of light. He produced a spectrum from sunlight and exposed sensitive thermometers to each color of light. He found that the heating effect increased towards the red end of the spectrum and kept increasing beyond the red light. Herschel realized that there had to be radiation in the solar light which was invisible. He called this portion of the solar spectrum *infrared.*

Based on these discoveries, the solar spectrum was divided into three subranges. Each subrange was characterized by wavelengths, the distances between two adjacent wave crests or troughs. A wave-length was measured in meters (m) or related units micrometers (µm). One micrometer is one millionth part of a meter: $1\ \mu m = 1/1{,}000{,}000\ m = 10^{-6}\ m$.

Visible radiation is contained in the range from 0.4 µm (violet light) to 0.7 µm (red light). Our eyes are sensitive to radiation in this portion of the spectrum. *Ultraviolet* radiation is characterized by wave-lengths shorter than 0.4 µm, while *infrared* radiation is characterized by wavelengths longer than 0.7 μm. Neither ultraviolet nor infrared radiation can be seen by humans.

Human eye. Light enters the human eye through the *pupil* and is focused by the *crystalline lens* on a light-sensitive tissue called the *retina*. Light impulses stimulate a layer of nerve receptors called *rods* and *cones*, which send signals to the brain. The shapes of these cells are explained by their names. There are more rods (about 120 million) than cones (about 7 million) in a single eye. All nerve fibers collect at the back of the eye to form the optic nerve. At the specific point where this large nerve leaves the eye, there are no rods or cones; this point is known as the *blind spot*. The rods contain a chemical substance, *pigment* (purpura or rhodopsin), which is very sensitive to the amount of light received by the retina, but cannot be used in color vision. This allows us to see only achromatic (white and black) images. Since rods do not require much light to react, this explains why we can see in the darkness and also why we cannot see colors well at night. In very bright light the pigment decolorizes completely, and rods are blind. The process is quickly reversed in the dark. The cones need much brighter light to work well. Each cone contains a colored, sensitive-to-light pigment. There are three kinds of cone cells, each one responding to one of the three primary colors: blue, green, and red. Our brain senses a range of colors by mixing the signals coming from each type of cone.

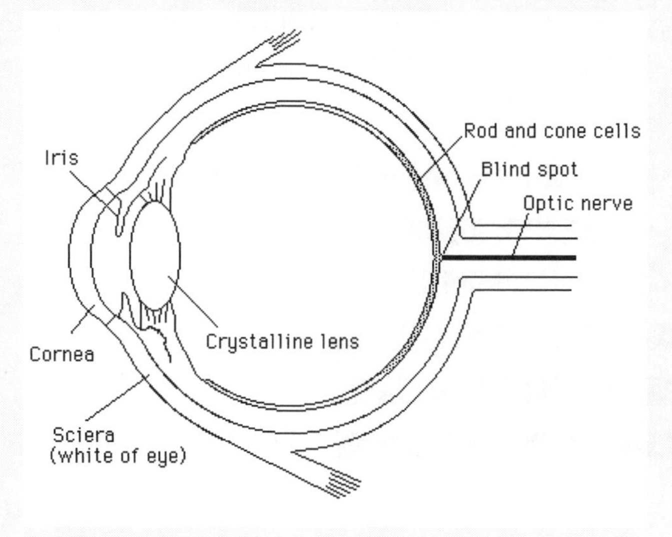

Solar radiation. Solar radiation is produced through a complex thermonuclear reaction. In this reaction, protons (hydrogen nuclei) are converted into alpha particles (helium nuclei). In the process, mass is converted to energy. Part of this energy is sent toward the Earth in the form of visible and invisible light. The Sun emits 44% of its energy as visible radiation, 7% as ultraviolet, and 49% as infrared radiation. Charged particles are also emitted by the Sun, and called the *solar wind*.

5.16. AFTER-IMAGE

Procedure: Color the drawing below by using two colored pencils, yellow and green. The area covered by stars should be colored yellow. The white stripes should be colored green. Stare at the lower right corner of the yellow rectangle for about 30 seconds. Then shift your gaze to a blank white space (covering the picture of the flag with your hand). You should see the red, white and blue of the American flag for a few seconds. Blinking may cause the image to last a little longer.

Comment: The change of color is due to the fact that cones respond to a specific color. After continued stimulation they become fatigued. When neutral light enters the eye, the difference in response between the fatigued cones and those which are at full potential, results in the seeing of complementary colors.

MIXING OF COLORS. Herman von Helmholtz (1821-1894) pointed out that there was a difference between color in light and color in paint. Color in paint is made from *pigments,* which are naturally colored powders formed by grinding up certain materials such as clays, rocks, plants, etc. When light falls on a colored object, part of it is *absorbed* and part of it is *reflected*. The portion which is reflected forms colors that our eyes can see. The color white is seen when all colors are reflected and mixed together. The color black is seen when no light is reflected.

5.17. COLOR BLACK

Materials: cardboard box, black paint.

Procedure: Paint the outside of the cardboard box black. Make a small hole in the front of it. Notice that the hole looks darker than the black surface of the box. Experiment with different sized boxes and both black or white interiors.

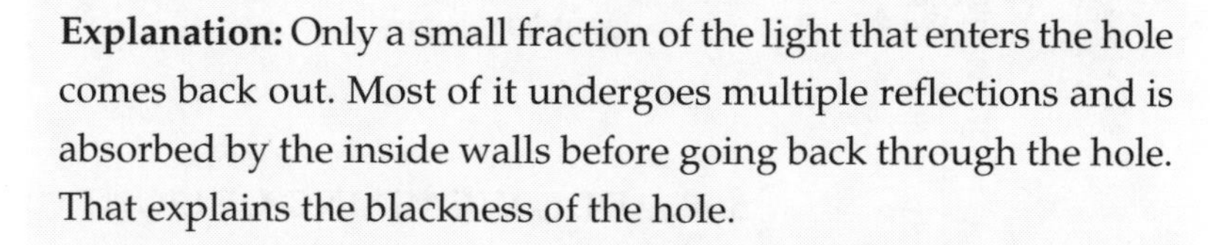

Explanation: Only a small fraction of the light that enters the hole comes back out. Most of it undergoes multiple reflections and is absorbed by the inside walls before going back through the hole. That explains the blackness of the hole.

Comment: We see the colors of various objects due to selective absorption. For example, the *chlorophyll* in green leaves absorbs much of the incidental light and allows only the green light to escape. As a result, the leaf appears green. When the same leaf is illuminated in a darkened room by any color other than green, it appears black.

White light can be made by mixing the three *primary* colors of light, *blue, green* and *red*:

white = blue ⊕ green ⊕ red

where the symbol ⊕ indicates the mixing of light. Mixing *red* and *green* yields *yellow*

yellow = green ⊕ red

Adding *blue* to *yellow* produces *white* light. That is why *blue* and *yellow* are called *complementary colors*. Similarly

magenta = blue ⊕ red
cyan = blue ⊕ green

The complementary color to *green* is *magenta*, a bluish red. The complementary color to *red* is *cyan*, a bluish green. Mixing each of the complementary colors produces the color white. This type of mixing is called *additive*.

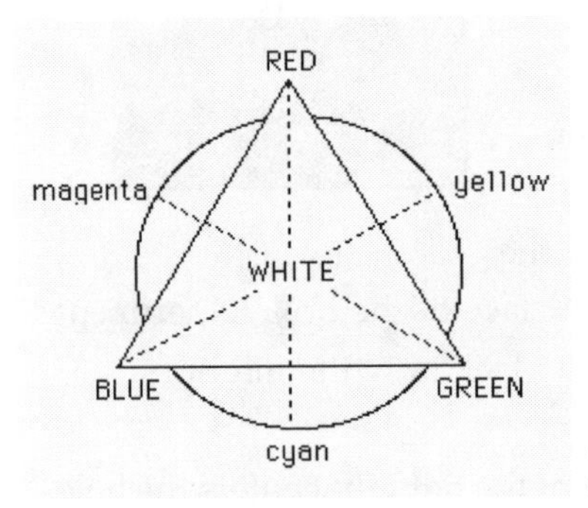

Primary and secondary colors

5.18. ADDITIVE MIXING OF LIGHT (1)

Materials: cardboard, spinning top, crayons.

Procedure: Make a cardboard circle 6 cm in diameter. Divide the cardboard into six sections. With the crayons, color each section in the following order: red, green, blue, red, green, blue. Fasten the wheel to the top. Spin the top. The colors will seemingly blend and produce a whitish color.

Explanation: By spinning the top, primary colors merge into a white one.

5.19. ADDITIVE MIXING OF LIGHT (2)

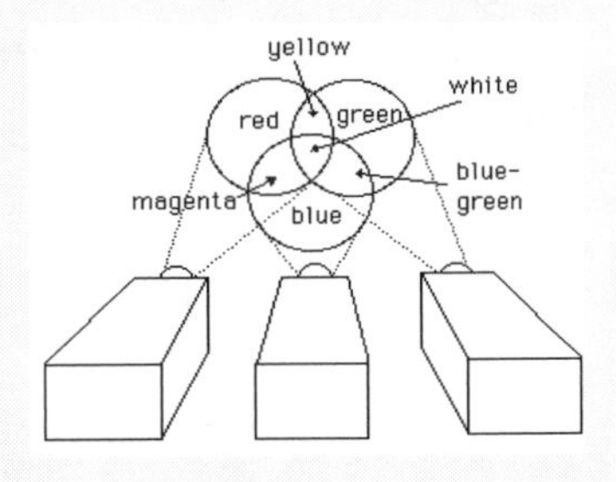

Materials: three slide projectors (or flashlights), pieces of red, dark blue and green cellophane.

Procedure: Cover the first projector with red cellophane, the second projector with dark blue cellophane and the third one with green cellophane. The colored light should have equal intensity. Shine all three projectors at one spot on a piece of white cardboard. The spot will appear white.

Explanation: The primary colors merge and yield a white light.

The mixing of colors in painting is different. For example, the mixing (the symbol @) of *blue* and *yellow* produces *green:*

green = blue @ yellow

The explanation is that the blue pigment (nearly pure dye) appears blue because it absorbs almost all the component colors of white light, except *blue*, a little of *green* and *indigo* (dark blue), which are reflected. A little of the green and indigo are also included because the pigments are not exactly pure colors. Yellow pigment reflects *yellow light*, as well as a little of the *green* and *red*. When yellow pigment is mixed with blue, *indigo* is absorbed by the yellow pigment, and *red* is absorbed by the blue pigment. The green is not absorbed at all but reflected by both pigments, which in turn produces a green effect. This type of mixing is called *subtractive.*

5.20. MIXING OF COMPLEMENTARY COLORS OF LIGHT

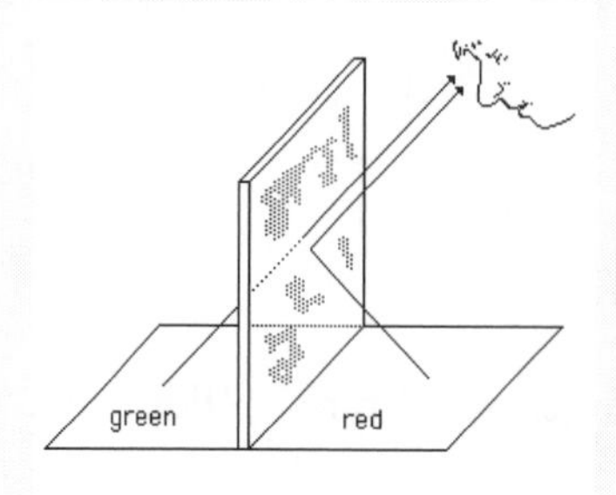

Materials: two complementary color squares (for example cyan and red), glass plate.

Procedure: Place two squares close together and put a glass plate upright between them. Position your eye so that one of the squares is visible through the glass and the other reflected in it. By varying the angle of the plate and thus changing the ratio of the light fluxes from the squares, you can completely decolorize the superimposed images. To achieve a white image, the colors must match perfectly.

THE BLUE SKY. Why is the sky blue? Isaac Newton argued that the blue color of the sky was due to the reflection of sunlight from hollow water droplets. His theory was completely wrong because such droplets do not occur in the atmosphere.

In 1881, John Tyndall (1820-1893) provided the correct explanation of this phenomenon. Based on research of Lord Rayleigh (John William Strutt, 1842-1919), Tyndall concluded that the blue color of the sky is due to the *scattering of light* (redistribution of light in all directions) by air molecules. Sizes of individual air molecules participating in *Rayleigh scattering* are smaller than wavelengths of visible light. The intensity (per unit-volume) of *Rayleigh scattering* is inversely proportional to the fourth power of the wavelength λ^4. From this one can calculate the relative scattering for various wavelengths:

Color:	violet	blue	green	yellow	orange	red
Wavelength λ [μm]	0.42	0.48	0.52	0.56	0.60	0.68
$1/\lambda^4$	32.1	18.8	13.7	10.2	7.7	4.7

The most intensely scattered are the shortest wavelength colors, such as violet and blue. Green is less scattered, and red is scattered very slightly. By mixing all of the colors in the proportions indicated in the last line of the above table (a lot of violet, less blue and green, and very little yellow, orange and red), the blue color of the sky is produced, as shown in the figure below:

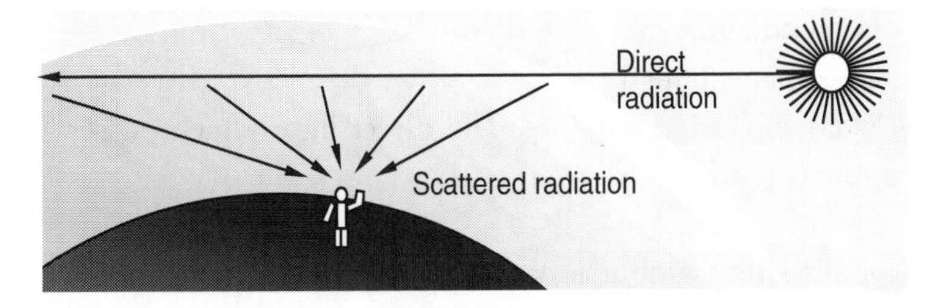

Scattering in the atmosphere

When you look at a distant object, you see light which is directly reflected by this object in your direction. In addition, you also see the scattered light coming from the same direction. As a result, the contrast of the distant object is reduced. The same object observed through a small tube made of paper looks sharper, because the tube reduces the amount of scattered light entering the eye.

Large particles (water droplets, ice crystals) do not obey Rayleigh's formula. Their scattering is called Mie scattering, in honor of the German physicist, Gustav Mie (1866-1957), who described it in 1908. Mie scattering is about the same for all wavelengths of visible light, and therefore appears white. Mie scattering explains the whiteness of thin clouds made of ice crystals (cirrus). Thicker clouds appear white because of Mie scattering by cloud droplets, but also because of multiple scattering by water vapor molecules, which are highly concentrated in clouds. Multiple scattering increases the number of ways in which beams of light of different wavelengths reach our eyes. Similarly, smoke above a cigar appears blue. But when it is puffed out of a mouth it seems whitish, because it is mixed with water droplets. Because of the multiple scattering, the sky close to the horizon looks whiter (atmospheric paths of light are longer there).

5.21. SCATTERING

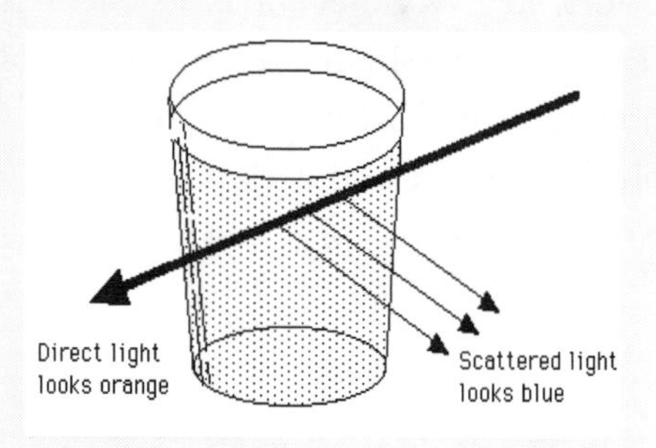

Materials: glass, milk, water.

Procedure: Fill a glass with water and add a drop of milk. Look at the side of the glass which is perpendicular to a direct beam. This side appears bluish. Look through the glass toward a window. The direct light which passes through the glass appears reddish.

Comment: Milk contains tiny globules of fat which scatter light (single Rayleigh scattering). When the concentration of milk in water is larger, the glass looks white because of multiple Rayleigh scattering.

RAINBOWS. In many old cultures rainbows were seen as bridges to heaven. In ancient Greece, they were associated with the goddess Iris, daughter of Thaumas and Electra. About the year 1310, a German Dominican monk, Theodoric of Freiberg, presented an explanation of rainbows which resembled a modern view of the subject. Based on experiments with prisms and transparent crystalline spheres, in his work, *De Iridae*, he argued that solar radiation entered an individual raindrop, changed its direction inside the droplet (by refraction), then underwent a reflection, only to be refracted once again before proceeding to the observer.

In the seventeenth century, a theory of rainbows was proposed by René Descartes (1600-1657). Descartes, in his famous work *Discourse on Method*, determined that when two refractions and one reflection were involved, a 42-degree angle resulted. Descartes used a drawing of a large droplet, as in the figure below:

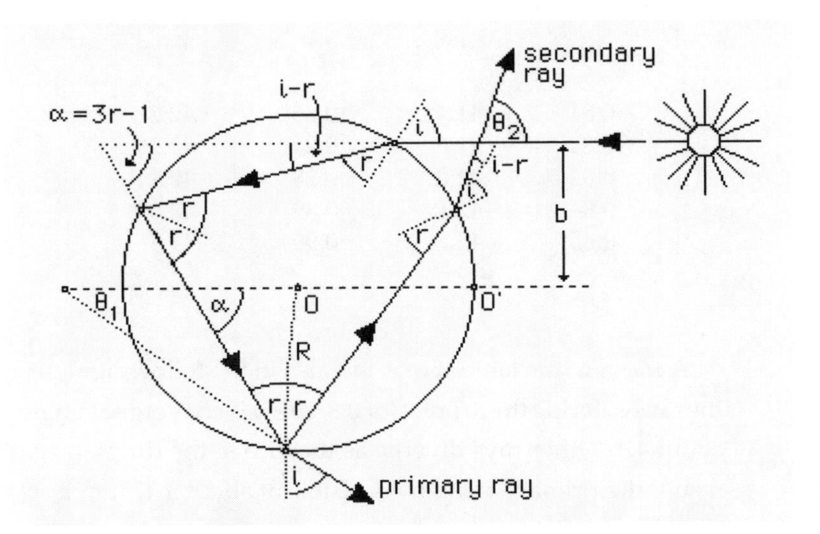

Descartes' calculation

From the figure, he found the angles Θ_1, Θ_2, which characterize the primary and secondary rainbows, are: $\Theta_1 = 4\mathbf{r} - 2\mathbf{i}$, and $\Theta_2 = 180° - (6\mathbf{r} - 2\mathbf{i})$, where **i** and **r** are incident and refraction angles (the secondary ray will be visible at the Earth's surface when the sun ray enters the drop below the O-O' axis). He arbitrarily assumed a drop unit radius R (e.g., 1 mm), so sin(**i**) = **b/R** = **b**, where **b** is the ray's distance from the drop axis. From Snell's law (page 110), he also obtained: sin(**r**) = **b/n**, where **n** is the index of refraction. Consequently, for a given values of **b** and **n**, he was able to find the angles **i** and **r**, as well as the angles Θ_1 and Θ_2. Repeating his calculation for **n** =1.332, we can obtain:

b	Θ_1	Θ_2	b	Θ_1	Θ_2
0.00	0.0°	180.0°	0.50	28.2°	107.7°
0.10	5.7°	165.7°	0.60	33.4°	93.1°
0.20	11.5°	151.3°	0.70	38.0°	78.6°
0.30	17.2°	136.8°	0.80	41.4°	64.8°
0.40	22.8°	122.3°	0.90	41.7°	53.3°

The obtained table indicates that for equal increments of b, increments of the angles Θ_1 and Θ_2 are not equal. Repeating the calculations for Θ_1 in the range $0.8 < b < 0.9$, gives:

b	Θ_1	b	Θ_1
0.81	41.4°	0.86	42.2°
0.82	41.6°	0.87	42.2°
0.83	42.0°	0.88	42.1°
0.84	42.1°	0.89	42.0 °
0.85	42.2°	0.90	41.7°

As seen in the table above and also in the following figure, the rays entering the droplet for $0.8 < \mathbf{b} < 0.9$, converge at angles about 42°. Other rays diverge as they leave the droplet. As a result, the primary rainbow is visible at about a 42° angle.

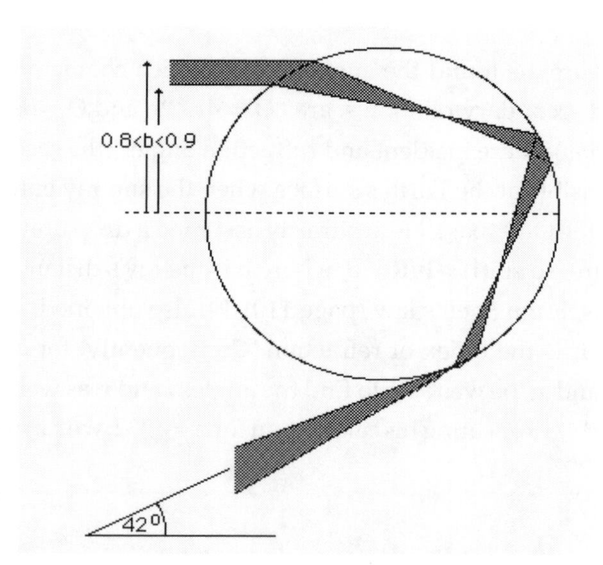

Explanation of a 42° rainbow

The inside of each raindrop acts like a mirror. Because only one reflection takes place inside a raindrop, a primary rainbow is very bright. A secondary rainbow is dimmer because it is caused by two reflections, as shown in the figure below:

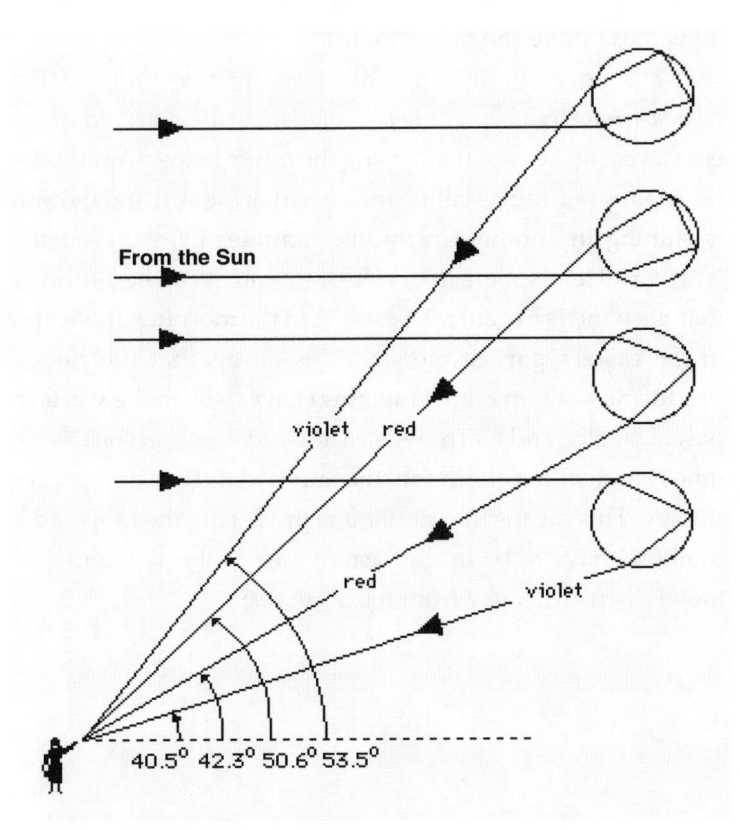

Primary and secondary rainbows

Differences in the index of refraction cause the outer edge of the primary rainbow to be red and the inner edge to be violet. The colors in the secondary rainbow are in the opposite order. This conclusion can be derived from the following table:

Color	Refractive Index	Max(Θ_1)	Min(Θ_2)
Violet	1.343	40.5°	53.5°
Green	1.335	41.8°	51.7°
Yellow	1.334	42.0°	51.3°
Red	1.332	42.3°	50.6°

where Max and Min indicate maximum and minimum values of the indicated angles (see the figure on the previous page) for all values of **b** in the range $0 < \mathbf{b} < 1$.

There is an old proverb: "*Rainbow in morning - sailor takes warning, rainbow at night - sailor's delight*". This old proverb can easily be explained using the figure below. A rainbow is visible when rain is falling in one part of the sky, and the Sun is shining in another. In middle latitudes (30°-60°), on the Northern Hemisphere, as a rule of thumb, it can be assumed that the wind is in general westerly. In the morning, the Sun is in the eastern part of the sky. This means that the clouds producing the rain (and a rainbow) have to be in the western part of the sky, and will move east, toward the observer. On the other hand, in the afternoon, the Sun is in the western part of the sky. This means that the clouds producing the rain (and a rainbow) have to be in the eastern part of the sky, and will move eastward, away from the observer.

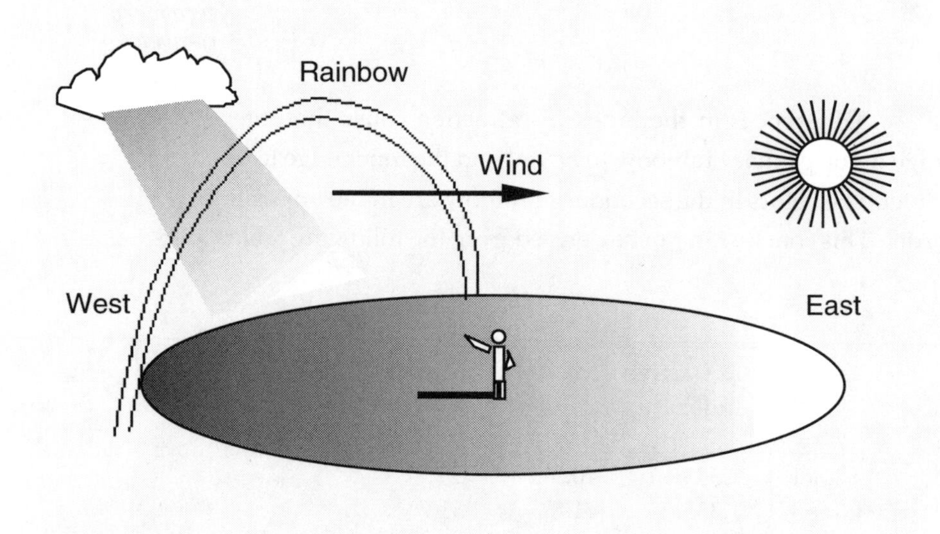

A morning rainbow

5.22. ANALYZING RAINBOW

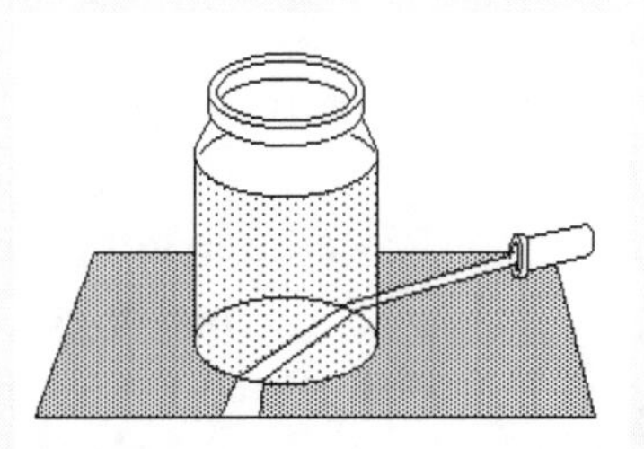

Materials: water-filled jar, flashlight, white paper.

Procedure: A rainbow-like refraction can be observed by shining a flashlight through the side of a water-filled jar (representing a rain drop) in a darkened room. For best results, place the jar on a piece of white paper. Aim the beam slightly downward to leave slight streaks on the paper. Starting with the beam next to the middle of the jar, slowly move the flashlight toward the edge of the jar without changing the direction of the beam. This causes the light to make larger and larger angles with the jar. At some point the light beam undergoes a total internal reflection (see Exp. 5.4), and emerges at an angle of 139° with respect to the light direction.

Comment: Notice that in this experiment, the light is directed in one specific direction. Real rainbows are caused by light directed toward our eyes by millions of droplets from various directions.

5.23 MAKING A RAINBOW (1)

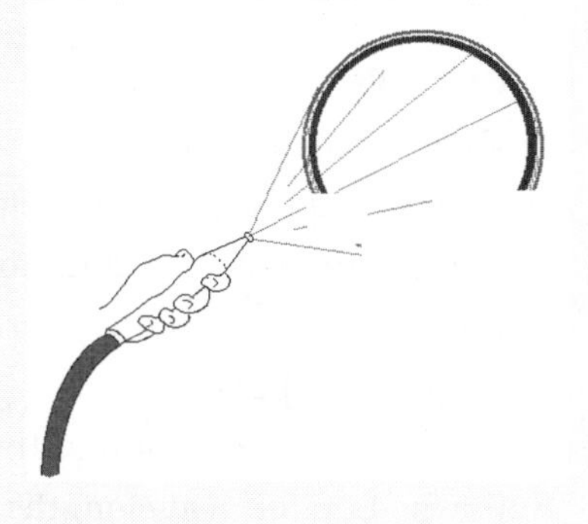

Materials: hose

Procedure: Spray water from a hose on a bright sunny day, with your back toward the Sun. The spray has to be below your eye. Observe a rainbow-like spectrum band in the stream of water. By adjusting the water spray (diameters of droplets), one can change the brightness and width of the rainbow.

5.24. MAKING A RAINBOW (2)

Materials: jar or bottle, water, white paper.

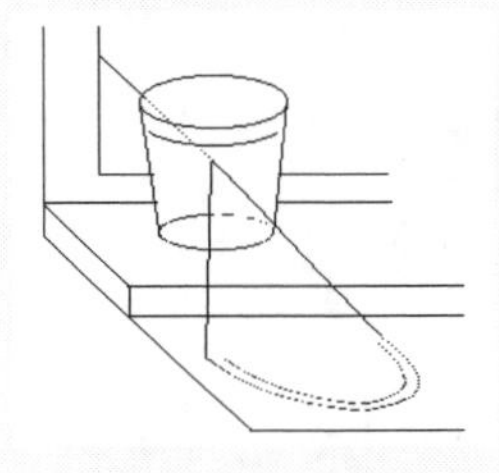

Procedure: Put a jar full of water over the inside edge of the window. Place a sheet of white paper on the floor. A rainbow-like spectrum band will become visible

5.25. MAKING A RAINBOW (3)

Materials: spherical glass flask, water, screen, flashlight.

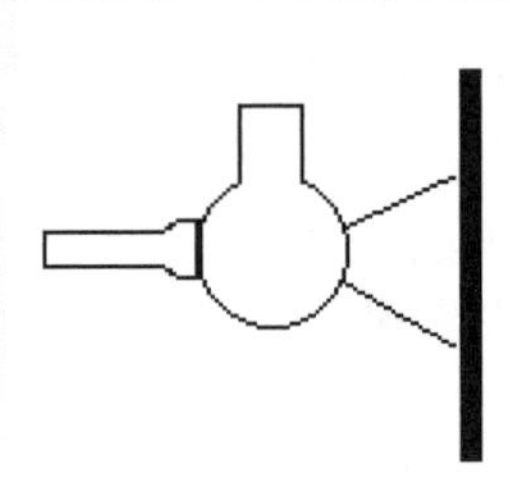

Procedure: Place a spherical glass flask filled with water in front of a flashlight. Illuminate the flask. A faint rainbow-like image will appear on a screen. It has the shape of a closed circle, with the color red inside, just opposite as in a real rainbow.

GREEN FLASH. Sometimes, a careful observer can spot a green light near the upper rim of the rising or setting Sun. This phenomenon is called the *green flash*. Have you ever seen the green flash? If you have, then according to an old Scottish legend, you will always be able to see into your own heart and be able to read the thoughts of others.

The phenomenon of the green flash can be explained as follows. Longer wavelengths of light are refracted less than shorter ones. Near the horizon, the difference in bending between red and blue is about 20' to 40', depending on the temperature lapse rate of the atmosphere. This causes one to see two Sun discs partially covering one another. The violet-to-green disc is a little higher in the sky, and the red one is a little lower. Most of the violet and blue rays are scattered away, as they pass through the thick atmosphere near the horizon, and are not seen. What can be seen is the green flash.

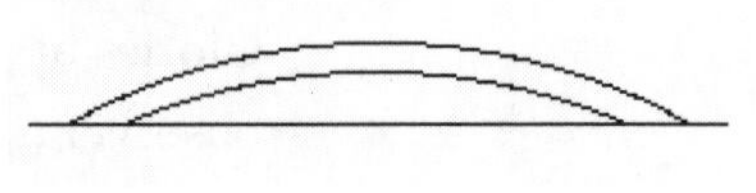

Sun discs

The longest green flash. Usually the green flash is visible for only a second. Richard E. Byrd (1888-1957), a well-known American explorer and organizer of the expedition to the South Pole, saw it for a much longer time. In 1934-1935, Byrd spent an entire cold season (summer on the Northern Hemisphere) alone in the advance camp. In September, when the Sun slowly rose above the horizon, marking the end of the cold season, he observed the *green flash* for 35 minutes.

5.26. GREEN FLASH

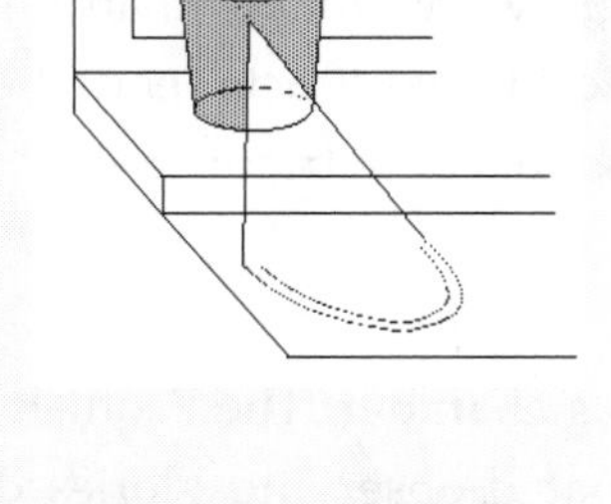

Materials: jar, water, milk.

Procedure: Place a jar full of water in bright sunlight over the inside edge of the window. Put a sheet of white paper on the floor. Add a few droplets of milk. As a result, the violet rim of a rainbow-like spectrum band will change its color to green.

Explanation: During the green flash phenomenon in the atmosphere, most of the violet and blue rays are scattered by air molecules. Similarly, in this experiment, the violet and blue rays are scattered by milk molecules.

GLORIES. Glories are still other optical phenomena in the atmosphere caused by refraction, reflection, and diffraction. To see a glory an observer must be situated in bright sunshine. Glories appear as colored rings around a shadow below an observer upon cloud decks or fog banks. If more observers are present, they can see only their own glories.

The glory

As in the case of primary rainbows, sunlight, which causes glories, is refracted and reflected back toward the Sun by cloud droplets. However, spherical droplets creating the glory are smaller than raindrops causing rainbows, and in addition diffract (bend) the incident and returning solar rays, so that both beams follow parallel paths.

Glories and Wilson's chamber. The Scottish physicist, Charles Wilson (1869-1959), became fascinated observing glories during his two-week visit to the meteorological observatory at the top of peak Ben Nevis in Scotland. What he usually saw was the shadow of the mountain on the surface of clouds, surrounded by two to five colored and complete rings. The rings appeared faint and diffuse, with the red outer edge. Wilson attempted to duplicate the effect in his cloud chamber (see Exp. 6.12). As a result, he discovered that in the absence of dust particles, water vapor can condense around ions in the air. Later, Wilson's cloud chamber allowed observations of the ions produced by X-ray radiation. For his research, Wilson received the Nobel Prize in physics in 1927.

No one so surely
pays its debt
as wet to cold
and cold to wet.
(Proverb)

CHAPTER SIX

MOISTURE

Water, Dew Point, Humidity, Hygrometers, Phase Changes, Condensation Nuclei, Coalescence, Cloud Electrification, Ice-Crystal Process, Forms of Precipitation, Cloud Classification, Rainmaking.

WATER. The ancient Greeks treated water as one of four basic elements. They had a fairly accurate understanding of the water cycle on the Earth. The Greeks knew that water ascended from the Earth's surface to form clouds in the atmosphere, and later returned to the Earth as precipitation. However, they did not fully understand why at some stage of the process water became invisible. To resolve this problem, Aristotle (384-322 B.C.) assumed that ascending water turned into air and then could again turn into water to form drops of clouds.

Anaxagoras (500-428 B.C.), who lived in Miletus, came to understand that heat caused air to rise. He also believed that atmospheric air was warm near the surface, cool above it, and that clouds resulted when rising air cooled.

In the seventeenth century, the French philosopher René Descartes (1596-1650), in the appendix to his famed *Discourse on Method,* correctly stated that air and water were two different elements. Descartes also suggested (incorrectly) that water was composed of smooth, slippery particles, like little eels, which joined and twisted around each other, but were not hooked together. He thought that water could change from solid to liquid or to gas, depending on how tightly its particles were knotted together. He concluded that evaporation occurred when water particles were strongly agitated, for instance, by the Sun. As a result, the particles could be separated easily from each other, fly into the air and form invisible water vapor. He also noted that even when water became invisible, its particles maintained their shape, distinct from that of the particles of air.

Evaporation as envisioned by Descartes

Some ancient philosophers believed that water vapor was composed of water and fire (or heat) and had no weight. Another belief was that water could be partly changed into earth. This believe came from observing that solid residue was usually obtained after the evaporation of water. Many scholars supported this idea, among them Johan Baptista van Helmont (1580-1644) and Robert Boyle (1627-1691). Boyle bolstered the above doctrine with the fact that the growth of plants could only be attributed to water. The blow to this theory came in 1769 when Antoine Lavoisier (1743-1794) found, through careful weighing, that the solid residue left by the evaporated water was silica, the earth material which had been dissolved from the glass, and was not the result of transmutation of water.

In 1666, the Italian priest Urbano d'Aviso concluded that water vapor consisted of hollow droplets of water filled with fire. Many scientists accepted this notion. One of them was the distinguished astronomer, Edmund Halley (1656-1742), later honored by having a comet named after him. Some individuals actually claimed that they saw the hollow droplets through a powerful lens. Among them was Horace Benedict de Saussure (1740-1799), who constructed the first human hair hygrometer.

In 1781, Joseph Priestley (1733-1804) began to experiment with mixtures of *inflammable air* (oxygen) and *dephlogisticated air* (hydrogen). He placed both gases in a closed vessel and passed an electric spark through it. The spark triggered a violent explosion and produced mist on the walls of the vessel. Priestley noted the mist, but did not make an effort to explain its presence. He informed Henry Cavendish, about his observations.

In 1783, Henry Cavendish (1731-1810), with the assistance of John Warltire, repeated the experiment. They observed that when two volumes of the *inflammable air* were mixed with one volume of the *dephlogisticated air*, all "airs" condensed into water, without any weight loss during the process. As a result, Cavendish concluded that water, long considered one of the four basic elements (see page 20) was not an element at all, because it consisted of hydrogen and oxygen. Similarly, air and earth were shown not to be basic elements. As a result, the theory of the "four elements" was finally rejected.

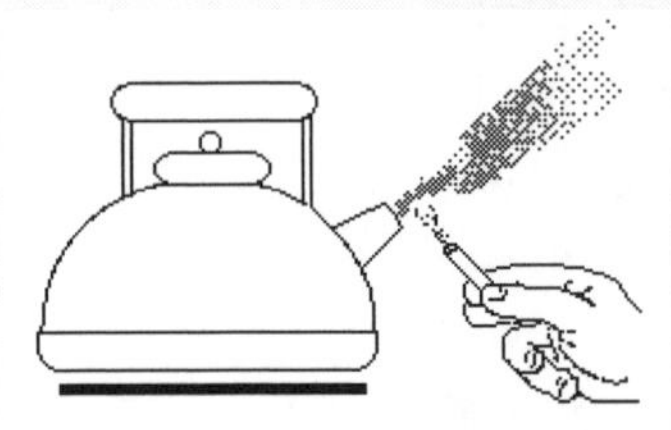

6.1. WATER VAPOR IS INVISIBLE

Materials: tea kettle, candle, hot plate, cigar.

Procedure: Bring some water to a boil in a tea kettle. Notice, that when the water is boiling, white steam emerges at a distance from the nozzle. Hold a lighted candle in the cloud of condensed steam. The cloud disappears. If cigar smoke is introduced, the clouds form more readily in the invisible portion of the stream.

Explanation: Water vapor is invisible and cannot be seen near the nozzle. At a distance from the nozzle the vapor cools and condenses in the form of a white cloud. The cloud consists of small droplets of water and therefore is visible. Heat from a candle is used to evaporate droplets. The cigar smoke provides particles on which condensation occurs.

DEW POINT. In 1751, Charles Le Roy (1726-1799), a professor of medicine at the Royal Academy of Sciences at Montpellier, France, made an important observation. Le Roy lowered the temperature of water in a glass container by adding ice. Consequently, he found that dew appeared on the outside of the glass as the glass cooled below a certain temperature. He called this temperature the *dew point temperature*. When the glass was heated, the dew disappeared. This process could be repeated whenever the glass container was heated or cooled.

Le Roy also studied the effects of various weather conditions. He discovered that air contained more water vapor on hot days than on cold ones. He concluded that a true measure of the air's humidity had to be relative. An actual mass of water vapor in the air had to be related to the maximum amount of water vapor which he could observe at a given temperature. Consequently, he formulated the concept of *relative humidity.*

Relative humidity ranges from 0 to 1. It is usually expressed in percentages (%), and in this case ranges from 0 to 100 %. When the air is completely dry, the relative humidity is zero. On the other hand, when the relative humidity is 100 %, the air is called *saturated.* During saturation, there is an equilibrium of evaporation and condensation rates. This equilibrium changes when temperature changes (the greater temperature, the greater energy of molecules in the liquid, and the greater evaporation). Using the concept of saturation, the relative humidity can be defined as a ratio of the actual mass of water vapor and the mass of vapor saturated at the actual temperature:

$$\text{Relative humidity} = \frac{\text{actual mass of water vapor}}{\text{mass of saturated water vapor}}$$

Le Roy believed that water vapor dissolved in air, just as salt dissolved in water. He also believed that water could reappear as a liquid in the same way salt could be precipitated out of a solution. Consequently, he was the first to use the word *precipitation* to describe rain, snow, dew and hail. Since further experiments revealed that water could evaporate into a vacuum, the assumption that water vapor dissolved in air was rejected. However, the word "precipitation" has remained in the language. Le Roy's theory was published in the famous French *Encyclopedie* (1751-1772), edited by Denis Diderot (1713-1784) and Jean le Ronde d'Alembert (1717-1783), and became widely known.

HUMIDITY. In 1802, the English chemist, John Dalton (1766-1844), known for his study on color blindness, realized that water vapor was a gas, one of many gases which mixed to form the atmospheric air. Dalton derived that the *relative humidity* could be equivalently expressed as a ratio of the actual vapor pressure to the equilibrium (saturation) water vapor pressure at the actual temperature:

$$\text{Relative humidity} = \frac{\text{actual pressure of water vapor}}{\text{pressure of saturated water vapor}}$$

There are many other characteristics of humidity. One of them is the *mixing ratio*, which can be defined as a ratio of the mass of water vapor in a given volume to the mass of dry air:

$$\text{Mixing ratio} = \frac{\text{mass of water vapor}}{\text{mass of dry air}}$$

In the saturation state, this quantity is called the *saturation mixing ratio.* The saturation mixing ratio over a flat liquid surface at sea level depends on temperature, as shown in the table below (the dependence is slightly different over curved, or over solid surfaces):

Temperature °C	Saturation Mixing Ratio g/kg	Temperature °C	Saturation Mixing Ratio g/kg
-40	0.10	5	5.42
-30	0.30	15	10.71
-20	0.78	20	14.91
-10	1.77	30	28.02
0	3.77	40	51.43

Relative humidity can change in two ways. First, the amount of water vapor in a given volume of unsaturated air can be increased by evaporation. When air reaches the saturation state, the evaporation and condensation rates are equal, and the relative humidity remains the same, equal to 100%.

Relative humidity can also be changed by varying the temperature. Assume that the actual content of water vapor is 5.42 g in 1 kg of air, at a temperature of 5°C. From the table above, it follows that because the equilibrium content of water vapor at this temperature is also 5.42 g/kg, the air is saturated, and its relative humidity is 5.42 g/kg : 5.42 g/kg = 100%. If the temperature increases to 15°C (the equilibrium content of water vapor at this temperature is 10.71 g/kg) the actual content of water vapor remains the same, but the relative humidity decreases to 50.6% (because 5.42 g/kg divided by 10.71 g/kg is 0.506 = 50.6%).

HYGROMETERS. The word *humid* comes from the Greek word *hygros*, which means "moist". An instrument which measures humidity is called a *hygrometer* (the Greek word *metron* means "to measure").

Cardinal Nicholas of Cusa (1401-1463) is credited for inventing the first instrument which measured humidity. The instrument, constructed around 1450, consisted of a balance of a large quantity of dry wool on one side and the equal weight of stones on the other. Nicholas of Cusa observed that the weight of the wool increased when the air was more humid and decreased when it was dryer. This type of hygrometer was still being used in the beginning of the eighteenth century.

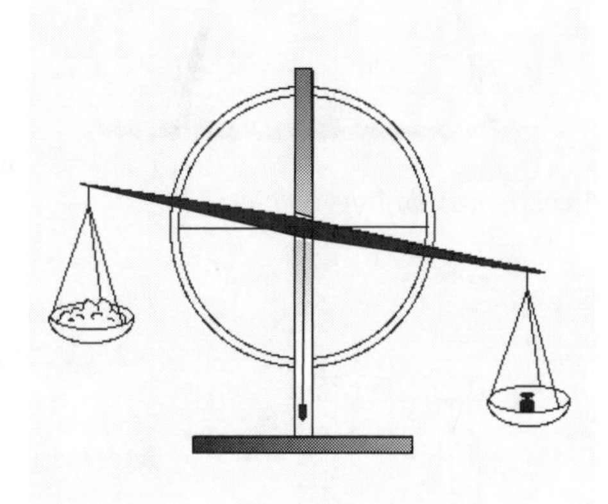

A balance hygrometer

In 1626, Santorre Santorio (1561-1636), a Venetian doctor, invented the *catgut hygroscope*. The catgut would twist or untwist depending upon the changes in the content of water vapor in the surrounding air. Robert Hooke (1635-1703) designed a *hair hygrometer* using hair from the beard of a wild goat as a sensor. The hair curled or uncurled, due to the changes in humidity, which was indicated by a pointer. The Accademia del Cimento of Florence used a *cord hygrometer*. It consisted of a rope about five meters in length, fastened at one end, passed horizontally to a pulley with a pointer, and stretched by a weight. In humid air, the cellulose fibers of the rope expanded decreasing its length, which was indicated by the pointer.

A cord hygrometer

In 1655, the Grand Duke of Tuscany, Ferdinand II, devised a *condensation hygrometer*. The instrument measured humidity by collecting drops of water condensed from air

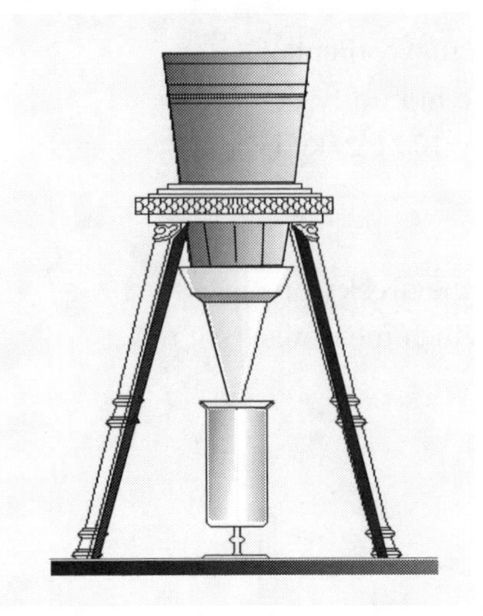
A condensation hygrometer

circulating around a funnel-shaped container filled with ice. As shown in the figure, the container was mounted on a tripod with its vertex downward. The greater the humidity in the surrounding air, the more the moisture condensed on the cold sides of the container. According to the original experimental notes, made on 27 August 1655, it was observed that the device, which was placed in the cellar of Ferdinand's palace, condensed 9 to 10 drops of water per minute. A second instrument, placed in the Great Hall, was able to produce 11 to 13 drops per minute.

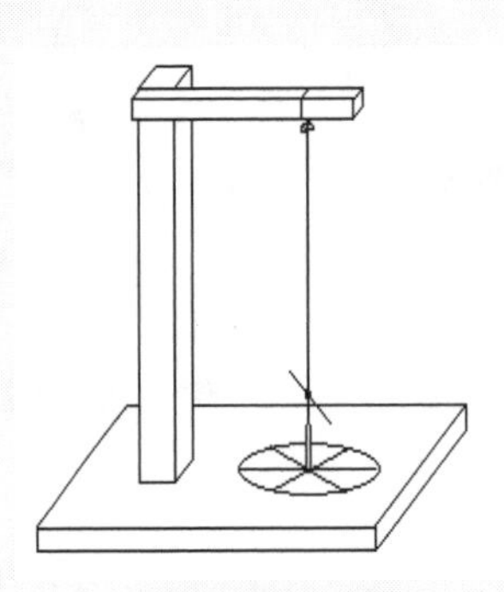

6.2. CATGUT HYGROMETER

Materials: catgut (ukelele string), simple wooden stand, rubber band, toothpick.

Procedure: Tie a gut string to the top of the stand. Attach the lower end of the string to the base using a rubber band. Glue a tooth-pick to the end of the gut. It will serve as an indicator. The gut string twists or untwists depending on the amount of moisture in the air.

Comment: In addition, a catgut hygrometer in the form of a popular "weather house" can be made from cardboard. Glue one end of the gut to the inner side of the roof's angle. Attach the other end to a horizontal twistable platform on which two figures are mounted. The direction of the twist of the gut can be found by trial and error.

In 1783, the Swiss scholar, Horace Benedict de Saussure (1740-1799), described a *human hair hygrometer* in his work *Essais sur l'Hygrometrie*. The hygrometer was based on the observation that an ordinary human hair, free of grease, would vary about one forty-second of its length (2.4%) between the two extremes: complete dryness or complete wetness (saturation).

6.3. HAIR HYGROMETER

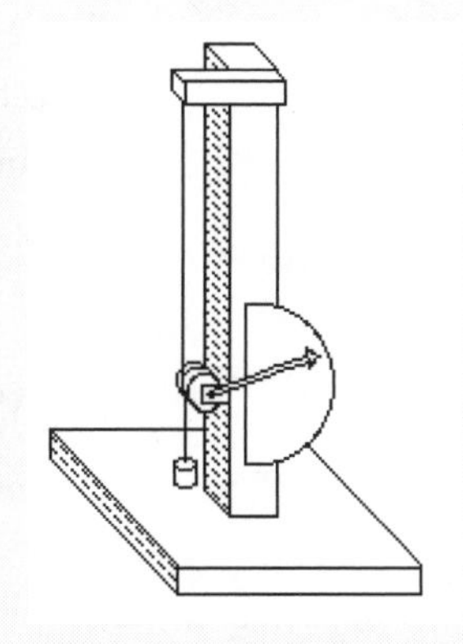

Materials: a few human hairs about 30 cm long, stand, spool, broom straw, nail, piece of tin, cardboard, 50 g weight.

Procedure: Wash the hairs in a shampoo solution and set aside to dry. Washing removes the natural oils, which would otherwise interfere with proper absorption of moisture. Attach one hair to the upper end of a stand and stretch it with a 50 g weight. Wind the hair two or three times around a spool (or a 1 cm long soda straw) attached to an axle (a long nail). The axle should be able to rotate in bearings made from a piece of tin and fastened to the stand. Fix a light pointer (a broom straw) to the axle and arrange a piece of cardboard as a scale. As moisture in the air varies, the hair will change its length.

Explanation: Hair tends to stretch as it absorbs moisture, and shrinks as it dries out. Blond hair is more sensitive to humid conditions and therefore works better. To calibrate, place the instrument above some warm water in a bucket and cover with a wet towel. When the pointer has moved as far as it will go, mark this point as 100% on the scale.

In 1755, William Cullen (1710-1790), a professor of medicine at Edinburgh, observed that a thermometer which was first immersed in alcohol showed a lower temperature. Cullen was the first to explain that this phenomenon was caused by cooling through evaporation. Later, the same principle was used to establish a standard method of obtaining humidity measurements with *wet and dry bulb thermometers*.

In August 1822, James Ivory (1765-1842) showed that the relative humidity can be measured with two, dry and wet bulb thermometers. He also derived an approximate formula for the relative humidity as a function of the actual temperature, the wet bulb temperature, and the air pressure. Three years later, the German physicist, Ernst Ferdinand August (1795-1842) obtained a similar result and named the dry-wet bulb hygrometer - a *psychrometer*, from the Greek words: *psychros*, meaning "cold", and *metron*, meaning "measure".

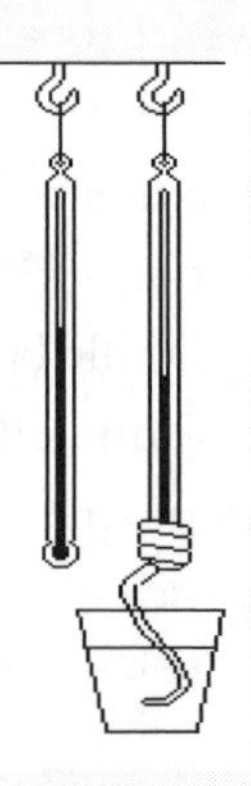

6.4. WET-DRY BULB HYGROMETER

Materials: Two inexpensive thermometers, white wick made of linen cloth, small bottle.

Procedure: Examine two thermometers by comparing their readings in different temperatures. If both agree, attach them to a piece of board, about 10 cm apart. Place a small bottle just under the right-hand side thermometer. Fasten a white wick made of linen cloth around the exposed bulb, and dip it into the bottle. Take the readings from both thermometers and use the table on page 157 to find the relative humidity. In the table, the temperature of the dry thermometer T is shown in the first column. The difference between the readings of both thermometers ($T-T_w$) is shown in the top row. Numbers in the table indicate the relative humidity.

Relative humidity (%) for an air pressure of 1000 mb

Dry-bulb temp °C	Dry-bulb - Wet-bulb difference , °C 0.5	1.0	1.5	2.0	2.5	3.0	3.5	4.0	4.5	5.0	10.0
-10.0	85	69	54	39	24	10	-	-	-	-	-
-7.5	87	73	60	48	35	22	10				
-5.0	88	77	66	54	43	32	21	11	0	-	-
-2.5	90	80	70	60	50	41	31	22	12	3	-
0.0	91	82	73	65	56	47	39	31	23	15	-
2.5	92	84	76	68	61	53	46	38	31	24	-
5.0	93	86	78	71	65	58	51	45	38	32	-
7.5	93	87	80	74	68	62	56	50	44	38	-
10.0	94	88	82	76	71	65	60	54	49	44	-
12.5	94	89	84	78	73	68	63	58	53	48	4
15.0	95	90	85	80	75	70	66	61	57	52	12
17.5	95	90	86	81	77	72	68	64	60	55	18
20.0	95	91	87	82	78	74	70	66	62	58	24
22.5	96	92	87	83	80	76	72	68	64	61	28
25.0	96	92	88	84	81	77	73	70	66	63	32
27.5	96	92	89	85	82	78	75	71	68	65	36
30.0	96	93	89	86	82	79	76	73	70	67	39

Dew-point hygrometer

In 1845, a professor of physics at the Collége de France in Paris, Henri Régnault (1810-1878), developed a *dew-point hygrometer*. The instrument was composed of a thin polished silver thimble ground to accurately fit a glass tube (left arm in the figure). The glass tube carried a small lateral tube connected with an aspirator. Some ether was poured into the tube. The air from the little tube was passed through the ether and produced bubbles. As a result, the ether was cooled by evaporation. In a minute, the temperature lowered sufficiently to produce an abundant deposit of dew. At this moment, the thermometer was examined in order to read the value of the dew point temperature.

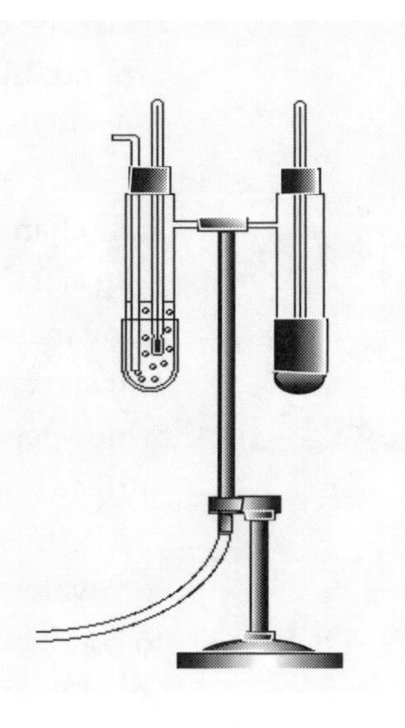

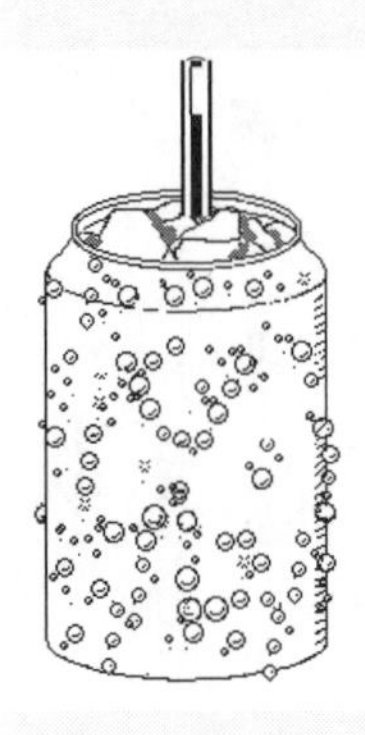

6.5. DEW POINT HYGROMETER

Materials: shiny tin can, thermometer, ice, newspaper, water.

Procedure: Place some water at room temperature, and a thermometer into a dry shiny can. Add some ice into the water and carefully stir it. Place the can on the printed page of a newspaper, so that the printing is clearly reflected from the can. Observe the outside of the can. Read the thermometer when dew begins to form on the outside of the can. The reading will be near the dew-point temperature.

PHASE CHANGES. Water is present in the atmosphere in a *solid* form as crystals of ice; in a *liquid* form as small droplets of clouds and rain; and in a *gaseous* form as water vapor. In the solid phase, the water molecules cannot overcome their electromagnetic attraction and tend to stay together. In the liquid phase the mutual attraction allows the molecules only to slide around. In the gaseous phase, the water molecules move so quickly that they almost break away from one another.

Evaporation is a process in which water in the liquid form changes into water vapor. The amount of heat (=2.50×10^6 J/kg at 0°C, and 2.26×10^6 J/kg at 100°C) required to evaporate a unit mass of water at constant temperature is called the *latent heat of evaporation*. Evaporation depends on the air humidity, wind speed (clothes dry faster in the wind), and evaporating surface (the larger the surface the greater the evaporation).

Condensation is a process in which water vapor changes into water. The amount of heat (=2.5×10^6 J/kg at 0°C) released to condense a unit mass of water at constant temperature is equal to the latent heat of evaporation.

6.6. EVAPORATION

Materials: container with water.

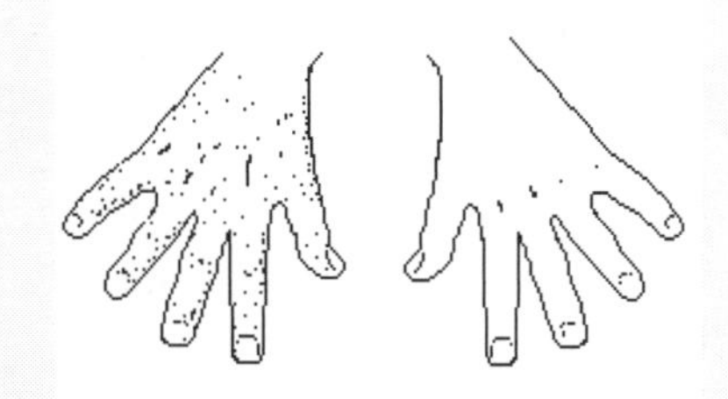

Procedure: Dip the palm of one hand in water, and keep the other one dry. Notice that the wet palm feels cooler. Blow some air over the wet palm. Notice that it feels even cooler.

Comment: Perspiration cools human body. During very humid weather, when relative humidity is near 100%, perspiration evaporates too slowly to cool the body. As a result, one feels hot, sticky and uncomfortable.

Freezing is a process in which water in the liquid form changes into ice. The amount of heat (= 3.34×10^5 J/kg at 0°C) released by a unit mass of water in order to freeze, at constant temperature, is called the *latent heat of freezing*. *Melting* is a process in which ice changes into water. The amount of heat (= 3.34×10^5 J/kg at 0°C) required to melt a unit mass of ice, at a constant temperature, is equal to the *latent heat of freezing*. Ice can be directly changed into water vapor by a process called *sublimation*. The reverse process is called *deposition*. The heat of sublimation is equal to the sum of the latent heats of vaporization and melting (2.83×10^6 at 0°C).

Phase changes

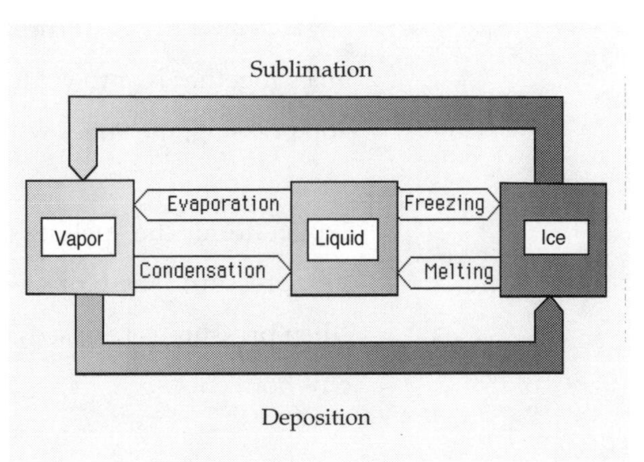

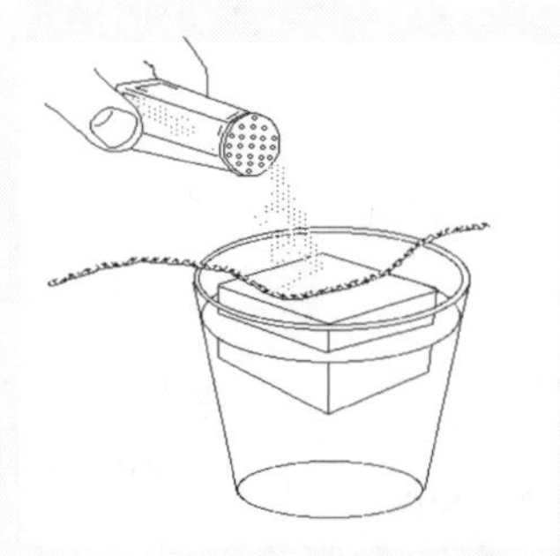

6.7. FREEZING (1)

Materials: string, ice cube, salt, water.

Procedure: Dip a string in water. Lay it across an ice cube. Sprinkle a little salt along the string. After few minutes the string freezes to the ice cube.

Explanation: The mixture of water and salt freezes at -16 °C. Consequently, when salt strikes ice, it lowers the freezing point below 0°C and causes it to melt. At the surface, the melted water releases its heat. This heat (latent heat of melting) is absorbed by the rest of the ice cube, which causes the water to freeze again.

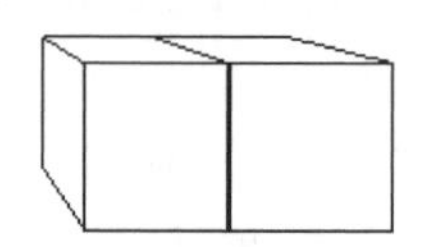

6.8. FREEZING (2)

Materials: ice cube, towel.

Procedure: Tightly squeeze two ice cubes together in a towel for several minutes. When you stop pressing, the cubes will freeze together.

Explanation: The melting point temperature decreases with pressure. Therefore, when you press, the ice melts. When pressure is released, the melted water freezes and joins both cubes.

6.9. DEPOSITION

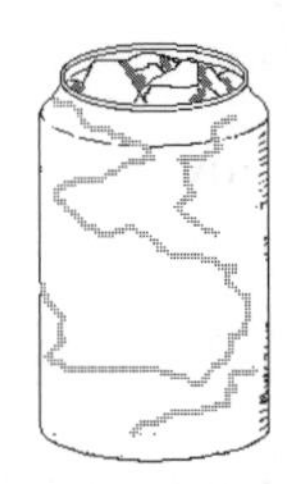

Materials: large tin can, thermometer, ice, salt, stick.

Procedure: Pack the can with alternate layers of ice and salt. Poke the mixture with a stick while you are packing it. After a while, the outside of the can will have a sub-freezing temperature. When the atmospheric humidity is sufficiently low (e.g., below 20% at 20°C), the saturation will occur below 0°C. At the temperature called the *frost point*, a delicate layer of frost (called white or hoar frost) will form resulting from a direct transformation of water vapor into ice. When humidity is high, the water vapor will condensate initially, and then freeze with the frozen dew not being the same as hoar frost.

6.10. FREEZING (3)

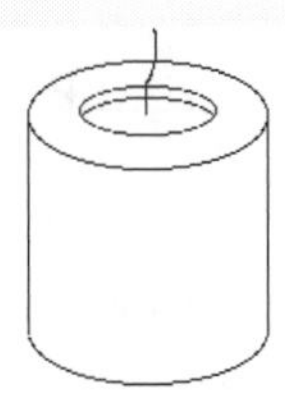

Materials: candle.

Procedure: Ignite the candle. After a few minutes extinguish it. Notice that a dip forms after the wax solidifies (freezes).

Explanation: Wax contracts as it freezes.

Comment: Water expands as it freezes. Therefore, ice is not as dense as water (it floats on the water's surface).

6.11. DRINKING BIRD

Materials: drinking bird toy.

Procedure: Observe the cyclical movements of the bird.

Explanation: The toy consists of a glass tube connecting the bird's head to the base. There is a liquid (ether) in the base. The water evaporates from the hat on the head and cools the vapor inside the head. This lowers the vapor pressure in the head. The higher pressure at the base pushes liquid up the tube. As a result, the replaced liquid makes the head of the bird tilt forward and dunk its head in the water glass. When the tube is almost horizontal, the pressure in the head and base, equalizes. Consequently, the bird returns to its up-right position.

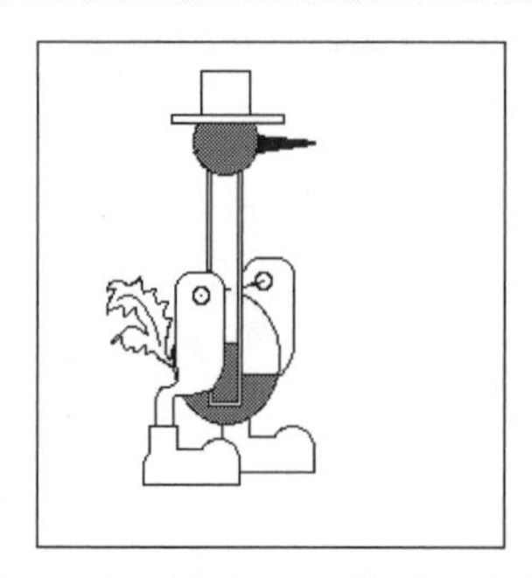

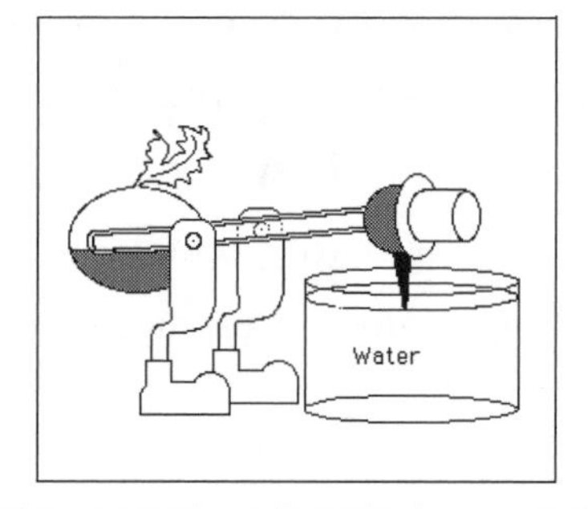

CONDENSATION NUCLEI. In 1875, the French scientist, Paul-Jean Coulier (1824-1890), tested the responses of fog to pressure changes in a totally enclosed system. He was surprised to observe that after a number of successful experiments in which he produced fog, something suddenly went wrong. He was not able to create fog, even though all the requirements of temperature, pressure and humidity were met. However, when he let some atmospheric air enter the container, fog could be formed again. He correctly concluded that fog could not be formed without the presence of dust in the air. These dust particles are presently called *condensation nuclei*. By repeating his experiments, gradually more and more *condensation nuclei* were removed from the air inside the container, preventing the formation of fog.

Scottish physicist, John Aitken (1839-1911), identified the most important kinds of condensation nuclei. Such nuclei are very minute and invisible to the human eye. They consist of particles such as volcanic dust, residues of combustion, and fine salt-dust (spray from the ocean after droplets dry). The smallest particles (10^{-7}-10^{-5} cm) which are present in the atmosphere are now called *Aitken nuclei*. *Large nuclei* measure 10^{-5} - 10^{-4} cm, while *giant nuclei* are larger than 10^{-4} cm.

6.12. CONDENSATION NUCLEI (1)

Materials: bottle, stopper, plastic tube.

Procedure: Place a stopper with a long tube into a bottle. Add a thin layer of water into the bottle. Shake the bottle vigorously. Put the free end of the tube into your mouth, and force the air into the bottle (a bicycle pump can also be used). Next, let the air go out quickly. The pressure in the bottle decreases and permits adiabatic cooling of the air (see page 81). To see its effect, insert condensation nuclei into the bottle. Take a breath of air from the bottle, and pinch the plastic tube. Hold a lighted match near the opening and release the tube. Allow the smoke to enter the bottle. Force the air into the bottle again. You will find that a much denser fog is being produced than before the smoke particles were admitted.

Explanation: There are three basic requirements for the formation of fogs or clouds: (1) water vapor in the air, (2) cooling to the point of saturation, (3) the presence of condensation nuclei. Under such conditions the molecules of water vapor can begin to condense.

Comment: A similar apparatus was built in 1849 by the American meteorologist, James P. Espy (1785-1860), who called it "nephelescope".

6.13. CONDENSATION NUCLEI (2)

Materials: salt (or sand), glass, carbonated beverage.

Procedure: Sprinkle salt (or sand) into a glass of a carbonated beverage. Observe the resulting bubbles.

Explanation: A carbonated beverage contains dissolved supersaturated (more that it would be in normal conditions) carbon dioxide, CO_2. Because energy is required to create bubbles, they do not form spontaneously. Salt sprinkled into beer provides a great number of nucleation sites allowing bubble embryos to grow. Similarly in the atmosphere, were it not for the presence of condensation nuclei, clouds would not form, unless the relative humidity was higher than 400%.

COALESCENCE. It takes about one million cloud droplets of average size to make an average raindrop. If this were the case, then what brings tiny cloud droplets together into much larger raindrops or snowflakes?

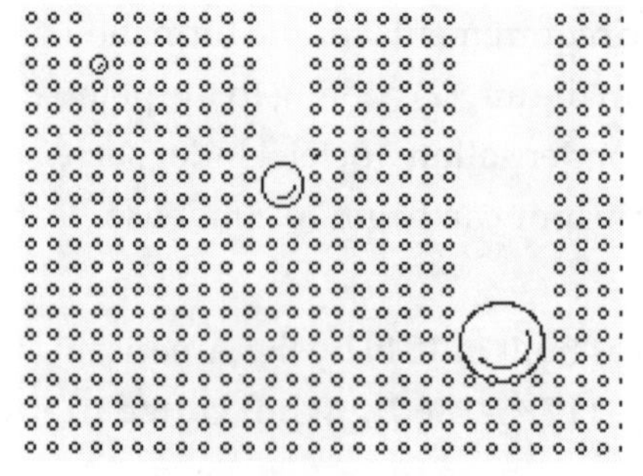

An idealized view of coalescence

Cloud droplets grow either by *coalescence* or by the *ice crystal process*. Let us first describe coalescence (the ice crystal process is discussed on page 170). During coalescence, large condensation nuclei which are always present in the atmosphere, encourage the growth of larger water droplets. Larger drops are able to overcome the lifting by rising warm air, and fall down. On their way down, they collide with smaller and slower falling droplets and consequently grow.

6.14. SALT ATTRACTS WATER

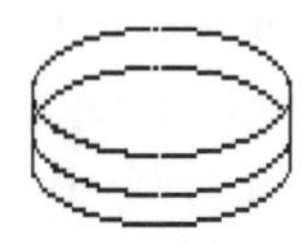
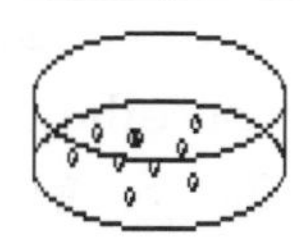

Materials: two small plates, water, salt.

Procedure: Pour some water in the first plate, and put several crystals of salt in the second one. After a few hours you will observe water droplets over the salt crystals. The effect can be hastened by covering both plates with a plastic sheet.

Comment: The relative humidity necessary for the onset of condensation is lower than 100% over droplets made of a salt solution. As a result, droplets can grow in unsaturated conditions.

CLOUD ELECTRIFICATION. The rupture of water drops in a strong vertical air jet produces considerable electrification. Large fragments of the broken drops acquire positive charges, while either the air ions or smaller droplets receive the compensating negative charge. Scientists believe that *coalescence* intensifies when the falling drops are electrically charged. Therefore, electricity in a cloud may encourage the growth of the cloud droplet to raindrop size.

Electric charges in the atmosphere can be observed during thunderstorms. The nature of thunder and lightning has been of interest to people for centuries. The first scientist to investigate these powerful phenomena was Benjamin Franklin (1706-1790), the American scholar and statesman, who participated in drafting the Declaration of Independence.

Electric charges. When one takes a shower, the splashing water produces negative charges in the bathroom. It is believed that negatively charged air creates a feeling of well-being and happiness. Negative ions of air are partly composed of oxygen, while positive ions are partly composed of carbon dioxide. Fresh air usually contains 700 to 1500 ions of oxygen in 1 cm^3. Forest air contains even more, about 15,000 ions of oxygen in 1 cm^3. There are regions where the amount of negative ions are larger than the amount of positive ions. The greater the number of negative ions, the better the air. Urban air is usually polluted, and therefore contains from 5 to 10 times smaller amount of negative ions than rural air. As a result, people living in large cities complain about losing some of their vital strength. The negative effect of positive ions on humans is caused by an increase of *serotonin*, one of the body's neurotransmitting substances. An excessive amount of serotonin is known to make a person nervous, irritable, and depressed.

In 1750, Benjamin Franklin proposed in a letter to the Royal Society that a long pointed iron rod, supported by an insulating glass stool, could be mounted in a small hut on a high chimney, or tower, to attract electricity from clouds. In June 1752, Franklin applied this idea by using a kite. The kite was made from a silk handkerchief and was attached to a string ending in an insulating silk ribbon. When a thunderstorm approached, Franklin observed that some of the fibres on the string stood erect. He placed his knuckle on the key attached to the string, and it produced a considerable spark. He was extremely lucky that he was holding the kite-string by an insulator. The experiment can cause death, and several other experimenters were electrocuted while attempting the same procedure.

Later, on 12 April 1753, Franklin collected electricity from an insulated rod and stored it in a Leyden jar. A Leyden jar (named after the Dutch city) was a metal-lined glass jar first constructed in the early 1700s, which allowed for the accumulation of static electricity. Franklin compared the obtained electricity with electricity of a known positive charge from a rubbed glass rod. On the basis of many similar experiments, he concluded that "*the clouds of a thundergust are most commonly in a negative state of electricity, but sometimes in a positive one, the latter, I believe, is rare*".

Clouds become electrified only after significant amounts of ice crystals and supercooled water droplets are presents (see page 170). Typically, there is a thin area of negative charges at the cloud top, and a thin area of positive charges at the cloud bottom. In the middle portion of the cloud, where water vapor molecules, water droplets, and ice crystals coexist at temperatures of about -15°C, there is a strong area of negative charge. Above this area, but much below the top, there is another area of positive charges. On the ground, below the cloud, positive charges build up. When the attraction between the positive charges on the Earth's surface (or another cloud) and the negative charges in the cloud overcomes the air's resistance, lightning flashes as *cloud-to-ground* (or *cloud-to-cloud*) lightning.

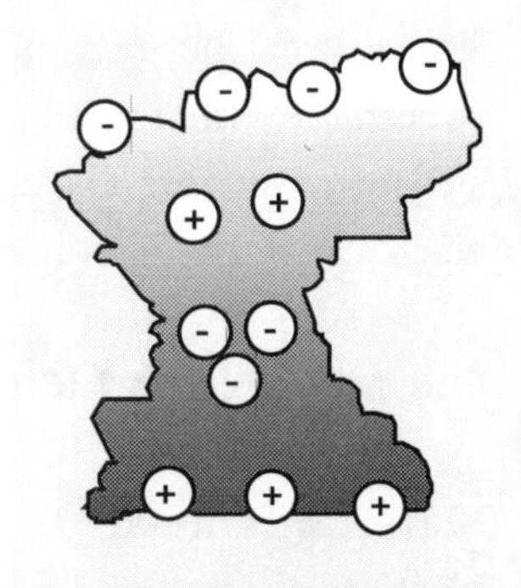

Distribution of charges in storm cloud

Benjamin Franklin (1706-1790)

Benjamin Franklin was born in Boston, Massachusetts, as the fifteenth child of seventeen. He was not only one of the founding fathers of the nation, but also a printer, writer, diplomat, and natural philosopher. He invented an improved stove, bifocal glasses, and a lightning rod. Franklin contributed to the science of electricity, and achieved a European reputation for his experiments with lightning. He worked out the courses of storms over North America, and was the first to study the Gulf Stream. He hypothesized that volcanic eruptions caused the unusually cool 1783 summer in Paris. Franklin represented the United States during the Revolutionary War at the Court of France. In 1900, Franklin was selected as one of the charter members of the Hall of Fame for Great Americans.

6.15. ELECTRIC CHARGES

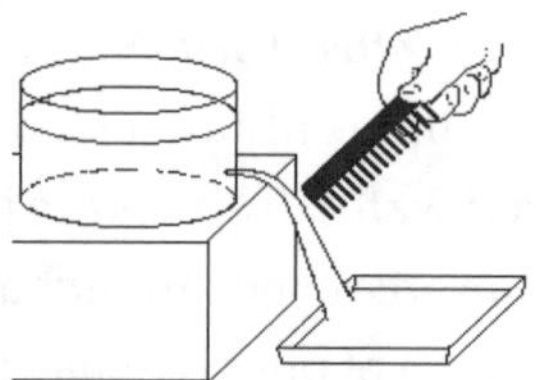

Materials: plastic comb, water source, wool cloth.

Procedure: Charge a comb by rubbing it with a wool cloth. Hold the comb near a thin stream of water. The stream will be attracted by the electrically charged comb and bend towards it.

6.16. MAKING A LIGHTNING DISCHARGE

Materials: metal dish, plastic putty, polyethylene sheet, tape.

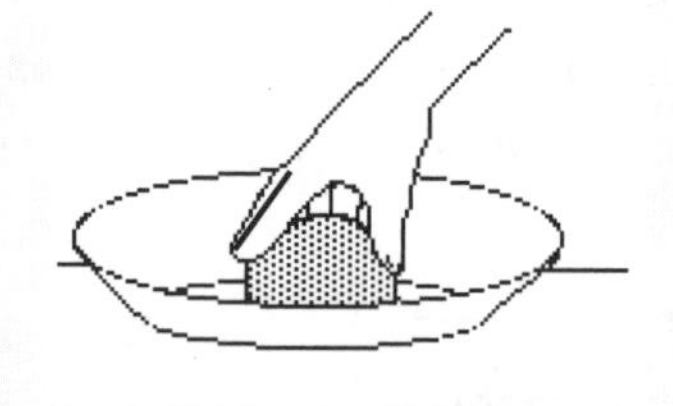

Procedure: Tape a polyethylene sheet to a table. Stick the plastic putty in the middle of the metal dish. Handling the dish through the plastic putty, rub the polyethylene sheet for about a minute. Place a metal object 2 - 3 mm from the plate. A bright spark will be visible in a darkened room.

Comment: Generated charges can be as high as a few thousands volts. Because the current is very small, the experiment is safe.

A *cloud-to-ground* lightning stroke generally consists of two stages. First, an avalanche of negatively charged electrons begins flowing down from the cloud in a forked pattern called a *stepped leader*. They advance in short steps, each about 1 microsecond long, with about 50-microsecond pauses.

In about 0.02 of a second, the stepped leader reaches the ground. At this time, a new stage, called the *return stroke,* zips upward. An intense (more than 1,000 times the current in home circuits) wave of positive charges flows upward in about 0.0001 of a second.

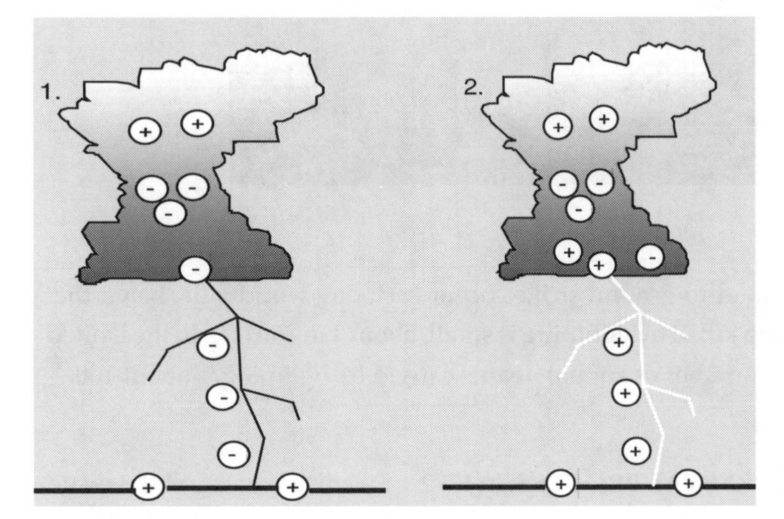

Stepped leader (1) and return stroke (2)

The *return stroke* produces light which can be seen as lightning, and heats air to 30,000°C (for comparison, the temperature of the Sun's surface is 6,000°C). This rapid heating causes the air to expand, and generates sound waves which are heard as thunder.

Distancing and photographing lightning. In order to find the distance to the flash after seeing it, one has to immediately start counting seconds until the thunder can be heard. The count multiplied by the speed of sound, which is 330 m/s (about 0.2 mi/s), tells the distance in meters to the flash. For instance, if the count was 5 seconds, the distance is: 5 s x 330 m/s = 1650 meters (1.02 mile).
In order to photograph lightning at night, pictures must be taken by using a camera (with a tripod) and setting exposures about one minute long, with a maximum lens opening. Because a storm usually lasts about 20 minutes, a 24-exposure roll of film can be used. The shutter should immediately be closed after each flash.

Thunderstorm

Up to 10 million cloud-to-ground strikes occur each day worldwide. Nevertheless, the risk of being killed by lightning is small, about 1 in 350,000. In the United States, the average number of annual deaths caused by lightning is about 100.

ICE-CRYSTAL PROCESS. In 1772, a Swedish scholar, M. Triewald, reported to the Royal Society of London that on a very cold day, 15 December 1731, he observed an unexpected phenomenon. When he touched a container of water, placed in the Palace of Nobility in Stockholm, the water instantly changed into ice. Triewald admitted that he was unable to account for what he witnessed. This unusual phenomenon was explained 180 years later by the German scientist, Alfred Wegener (1880-1930), who was also known for proposing that the Earth's continents were in motion.

In 1911, Wegener made two important observations. The first one was that water can be *supercooled,* which means that it can exist in a liquid form in temperatures way below the melting point (0°C). The second observation was that ice crystals in the air "pull" water vapor, because the vapor pressure near the ice surface is lower than it is near the surface of the water. By connecting these two ideas, he concluded that when ice crystals and water droplets are present in the air at the same time, the water droplets will evaporate. The water vapor which forms will move toward ice crystals and then condense. As a result, ice crystals will grow.

6.17. SUPERCOOLED LIQUID

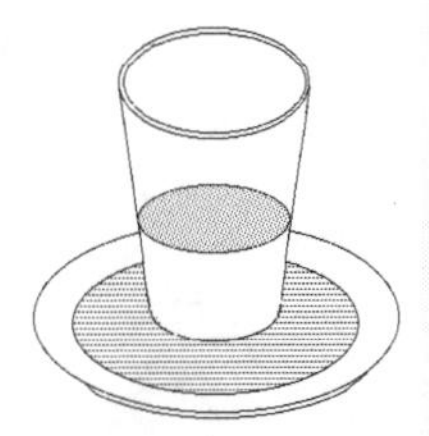

Materials: crystals of sodium thiosulfate $Na_2S_2O_3$ (low toxity substance used in photography, available as "hypo" in drug stores), glass, heat source.

Procedure: Fill a glass to the half mark with crystals of sodium thiosulfate. Heat the glass in a hot water bath, until the crystals melt to a transparent liquid. If there are any impurities on the glass bottom, filter them out using a funnel and cotton. Cover the glass, and allow the liquid to cool for about 15 minutes to a room temperature of about 20°C. Shake the glass. The liquid will immediately freeze. Touch the glass and notice how warm it feels.

Explanation: The freezing temperature of sodium thiosulfate is 48°C. In a room temperature of about 20°C the sodium thiosulfate is a supercooled liquid. Dropping a small crystal, or shaking the glass triggers the freezing process. During this process, the latent heat of freezing is released and warms the glass.

Comment: Bottles filled with liquid sodium thiosulfate and kept in coat pockets have been used by some as personal heaters in winter, because they release heat when shaken.

A decade later, Wegener's theory was confirmed through observations made by the Swedish meteorologist, Tor Bergeron (1891-1976), a member of the famous Bergen School of meteorology in Norway. In February of 1922, while vacationing near Oslo, Bergeron frequently walked along a narrow road on a hill in the fir forest. The road was parallel to the contours of the hillside and the hill was often covered by a supercooled stra-

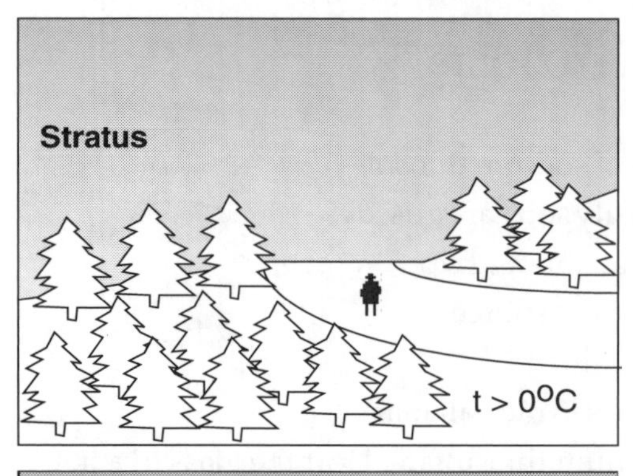

Bergeron's observations

tus-cloud layer. Bergeron noticed that the cloud did not enter the road tunnel when air temperatures were below freezing, but did enter when temperatures were above the freezing point. Recalling Wegener's theory, he concluded that ice on the trees had attracted water vapor, caused the evaporation of water droplets, and caused the cloud to vanish. Bergeron initiated intense research, and in 1933, developed his *ice-crystal theory of cloud formation*. His theory was later elaborated by the German physicist, Walter Findeisen (1909-1945), and is now widely accepted as the *Bergeron-Findeisen theory* of rain formation.

According to this theory, in order to create a crystalline structure, liquid water molecules must form in a certain order. Most droplets are so small that there is little chance that their molecules will achieve the necessary alignment, unless the temperature approaches -40° C. Below this temperature, all water molecules, in all phases, will freeze. Freezing will occur more readily if particular kinds of particles, called *freezing nuclei*, are present. Such particles (dust of certain clays) have a crystal structure similar to ice, and allow supercooled water to crystallize on contact.

Further growth of ice crystals occurs because the saturation vapor pressure over ice crystals is lower than over supercooled water. This means that when air is saturated with respect to liquid droplets, it is supersaturated with respect to ice crystals, i.e. its relative humidity exceeds 100% (for example, it is 121% at -10° C). If some crystals are present in the air, water molecules diffuse onto the ice and the process of *crystallization* accelerates, as shown in the following figure:

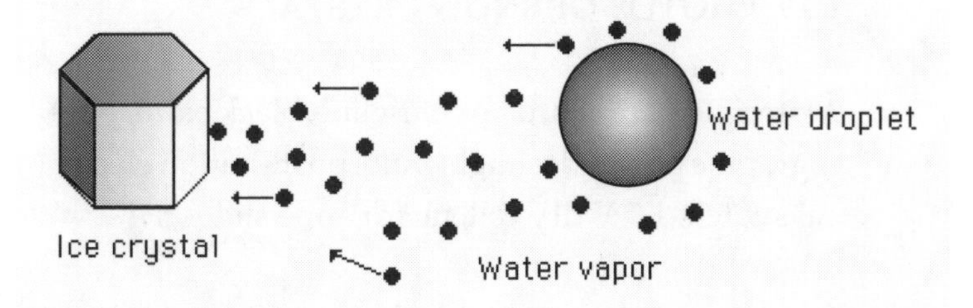

Ice-crystal process

6.18. NUCLEATION

Materials: sugar, water, ice, salt, glass.

Procedure: Place a saturated solution of sugar water into a glass. Supercool the glass in an ice and salt bath. Scratch some ice dust from an ice cube. When ice nuclei fall on the sugar solution, large sugar crystals form on the surface.

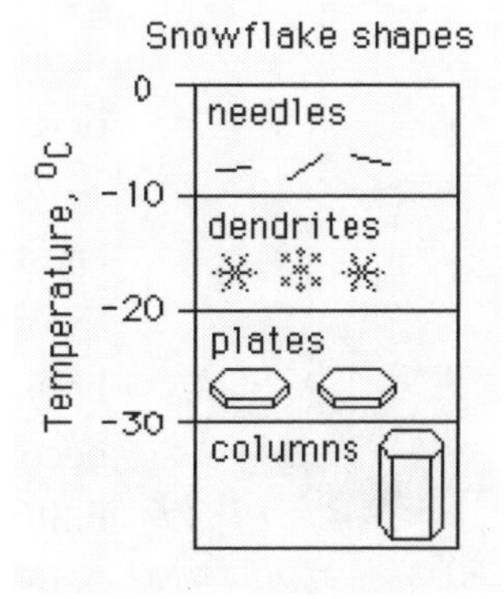

FORMS OF PRECIPITATION. In cold weather, falling ice crystals may arrive at the Earth's surface as aggregations in the form of snowflakes. As illustrated in the figure, shapes of snow flakes depend on the temperature in the cloud.

After leaving a cloud, flakes may pass through a layer of warm air below and melt, then go through cold air near the ground and refreeze. They finally reach the ground as *ice pellets* or frozen rain drops, also known as *sleet*. However, most often the crystals melt, stay melted, and come to Earth as *rain*. When supercooled drops of water collide with ice crystals, they freeze and form a coat of ice. This process is called *riming (or accretion)*. Riming in a cloud produces *graupel*, ice particles 2-5 mm in diameter.

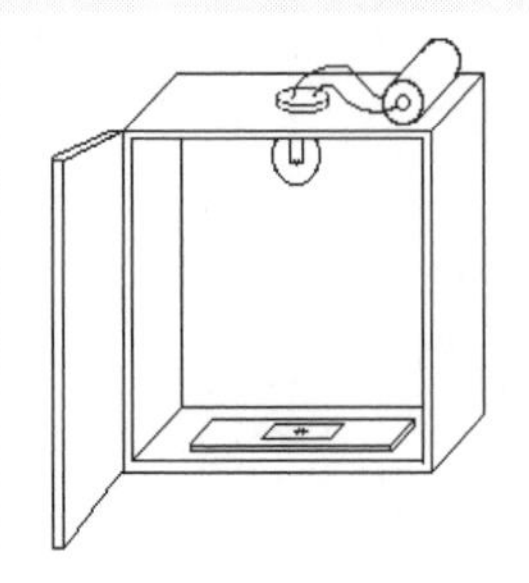

6.19. PHOTOS OF SNOW CRYSTALS

Materials: cardboard box, vaseline, black paint, photographic paper, flashlight bulb, push-button electrical switch, 1.5 V dry cell, microscope slides.

Procedure: Find a cardboard box, with sides of about 30 cm long. Paint the inside black. Place a flat holder for photographic paper on the bottom. Attach a flashlight bulb to the inside top face of the box, and connect it to a small push-button electrical switch and 1.5 V dry cell. Coat several standard microscope slides (3 by 10 cm) with Vaseline. These slides will be placed on top of the photographic paper, one at a time, when samples of snow crystals are obtained. Set the apparatus outdoors a few hours before use. Expose each plate until a single snow crystal or snowflake falls on it. Insert the plate into the box, placing it on top of the photographic paper. Close the box and press the switch to expose the film for three seconds. Open the box in the dark and remove the paper. Develop the photograph in a commercial shop. Your print will show a silhouette of the snow crystal with shades of light and dark (due to variations in light transmission through it).

When the air temperature at the Earth's surface is slightly below freezing, *freezing drizzle* (when drops are smaller than 0.5 mm) or *freezing rain* can occur. In this case, supercooled drops of drizzle or rain freeze upon striking cold objects.

Hail is a product of thunderclouds, and forms when ice crystals are carried alternatively by updrafts and downdrafts through layers of warm and cold air. By melting and refreezing, hailstones grow, collecting ice layer upon layer, until they become heavy and fall to Earth. The "history" of a single hailstone can be observed on a cross section made of concentric layers of clear (due to slow freezing) and milky ice (due to rapid freezing which traps air bubbles). The largest reported hailstone was 14 cm in diameter (0.76 kg in weight), and fell in Coffeyville, Kansas, in September 1970.

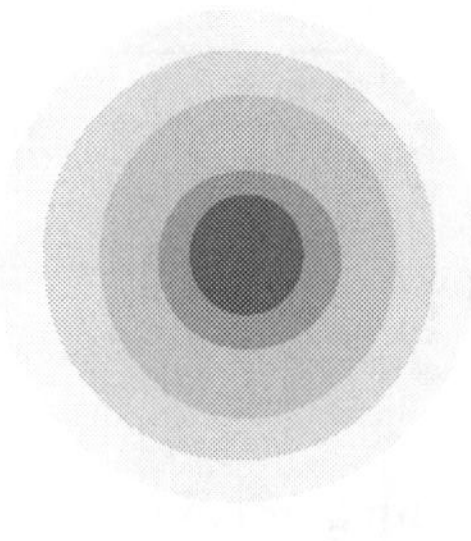

A cross-section of a hailstone

Flying through thunderstorm clouds. In 1930, a deadly demonstration of the hail formation process was provided by five glider pilots in Germany. The pilots soared into a thunderstorm in the region of the Rhön Mountains, and were trapped in the cloud's powerful drafts. After loosing control of their planes, all five parachuted out. Powerful updrafts immediately lifted them to regions of subfreezing temperatures. Then they fell down with the downdraft, only to be raised up and frozen again. After a series of lifts and falls they literally became human hailstones. When they finally reached the Earth, they were all frozen stiff and only one survived. Similarly in 1982, an Australian parachutist was trapped for half an hour in a cumulonimbus cloud. When he opened his parachute, he was lifted by an updraft from a height of 2 km to 4 km. The parachutist rescued himself by cutting out from his main parachute. As a result of free-falling, he was able to move through the cloud. About 500 m above the surface, he opened his reserve parachute and then landed safely on the ground.

Hail

In a popular artistic representation a raindrop looks like a teardrop. Actually, small raindrops (radius< 1mm) are spherical, larger ones assume shapes like that of a hamburger bun.

As rain droplets fall, they encounter frictional forces exerted by the air. The frictional force balances the gravity force and causes droplets to fall with a constant speed called the *terminal velocity*. As seen in the table below, the terminal velocity of droplets increases with their diameter. The table also indicates that if rain drops are not large enough, they might evaporate before reaching the Earth's surface. This type of precipitation is called *virga*.

Diameter cm	Terminal velocity m/s	Fall distance for complete evaporation, m	Type
0.00002	0.0000001	-	condensation nuclei
0.0020	0.01	1	typical cloud droplet
0.01	0.27	25	large cloud droplet
0.02	0.70	70	drizzle
0.1	4.0	400	small raindrop
0.2	6.5	650	typical raindrop
0.5	9.0	910	large raindrop

6.20. SNOW CRYSTALS

Materials: 1-3% solution of polyvinyl formal (powder) in ethylene dichloride (in the USA available e.g., from Polysciences, ph. 1-800-523-2575), adequate ventilation required because inhaling of ethylene dichloride might cause irritation of respiratory tract, glass microscope slide, eye-dropper.

Procedure: Immerse a clean glass microscope slide in the prepared solution (which should not be colder than -5°C) for about 30 seconds. Then expose it to falling snow. A captured crystal becomes submerged in the solution. It should be kept at a subfreezing temperature for few minutes until the solvent evaporates. This leaves the crystal encased in a thin, but tough, plastic shell. Large snowflakes can be additionally covered with a drop of a solution from an eye-dropper. In room temperature, the plastic membrane will retain the surface structure of the original crystal.

6.21. MODEL OF A SNOW CRYSTAL

Materials: a few magnified photo-copies of the drawing below.

Procedure: Cut the photo-copies out and glue them together in order to obtain the hexagonal rings found in snow flakes.

Comment: A water molecule has the shape of a teddy bear's head (negatively charged oxygen atom) with two ears (two positively charged hydrogen atoms). The angle between the center of the oxygen atom and the centers of two hydrogen atoms is 105°. The positioning of the atoms in the water molecule causes a characteristic arrangement of ice crystals in the form of hexagons (six-sided structures). In the model, the oxygen atom is represented by the body of the cube. Hydrogen atoms are attached to any two small triangular surfaces.

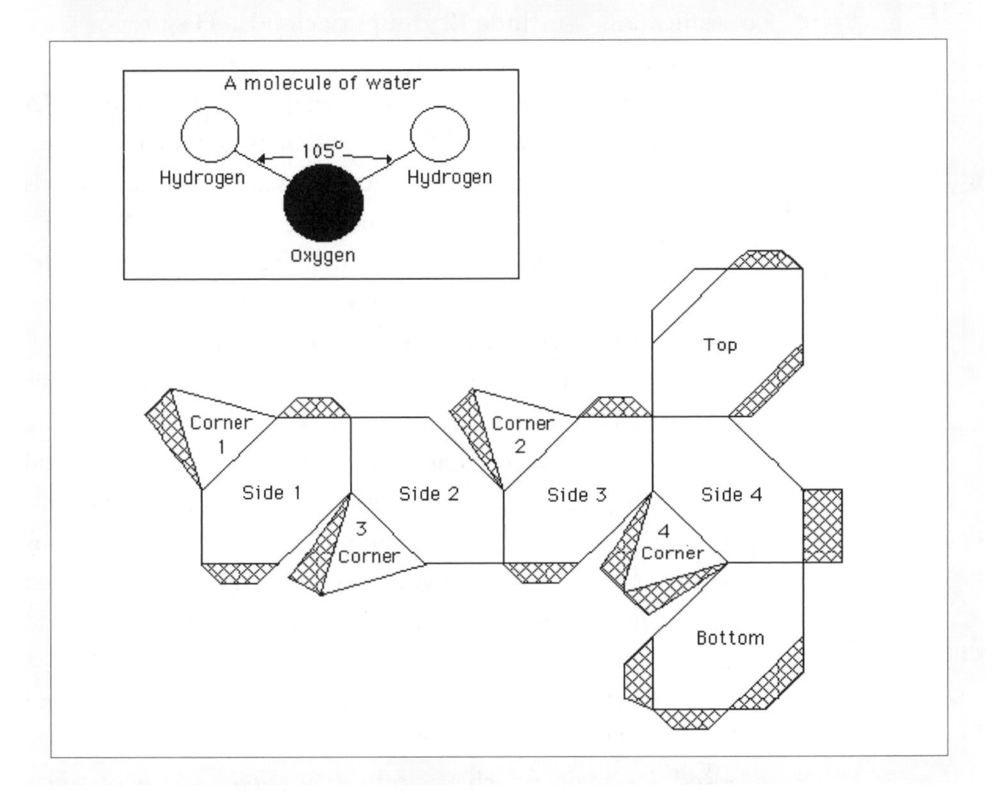

Rain gauges. The first known rain gauge was described in the manuscript *Arthastra* by the Indian author, Kautilya, in about 400 B.C. The rain gauge was simply a bowl with a diameter of about 45 cm. Rainfall measurements were taken regularly with this gauge, and were used to determine the annual crop to be sown. The first mention of a simple rain gauge in Europe was made by Benedetto Castelli in a letter to Galileo Galilei on 18 June 1639. A presently used rain gauge consists of a funnel collector, attached to a long tube. The depth of water in the tube is measured with a ruler in millimeters or inches. One millimeter of rain is equivalent to 1 liter of water per square meter of the Earth's surface.

CLOUD CLASSIFICATION. In 1802, the French naturalist, Jean Baptiste Lamarc (1744-1829), proposed the first classification of clouds. In the third volume of his *Meteorological Yearbook,* he introduced five forms of clouds: hazy, massed, dappled, broomlike and grouped. Three years later, he extended his classification to include 12 groups of clouds. His propositions did not receive wide acclaim.

In 1803, a British pharmacist, Luke Howard (1772-1864), published *On Modification of Clouds,* in which he introduced another cloud classification. Howard used Latin words to describe various types of clouds. He named sheetlike, thin layers of clouds *stratus,* from the Latin word meaning "widespread" or "layered". Puffy clouds were called *cumulus,* from the Latin word "heap". Wispy, high clouds were called *cirrus,* from the Latin "curl of hair". To denote clouds which produce showers, Howard used the Latin word *nimbus,* meaning "rainstorm". He also combined the terms to describe cloud combinations: *cirro-stratus, cirro-cumulus,* and *strato-cumulus.* Howard's classification became widely accepted. A famous German poet, Johann Wolfgang Goethe (1749-1832), dedicated four of his poems to Howard, in appreciation of his cloud classification.

In 1887, Britain's Ralph Abercromby (1842-1897) and Sweden's Hugo Hildebrand Hildebrandsson (1838-1920) proposed a system which basically has been used ever since.

Within the system, 10 principal cloud forms (genera) are divided into 4 primary cloud groups, dependent upon the height of their base above the surface. The four groups are: *low clouds* (the base height 0-2 km), *middle clouds* (the base height 2-8 km), *high clouds* (the base height 8-18 km), and the *vertically developed clouds.*

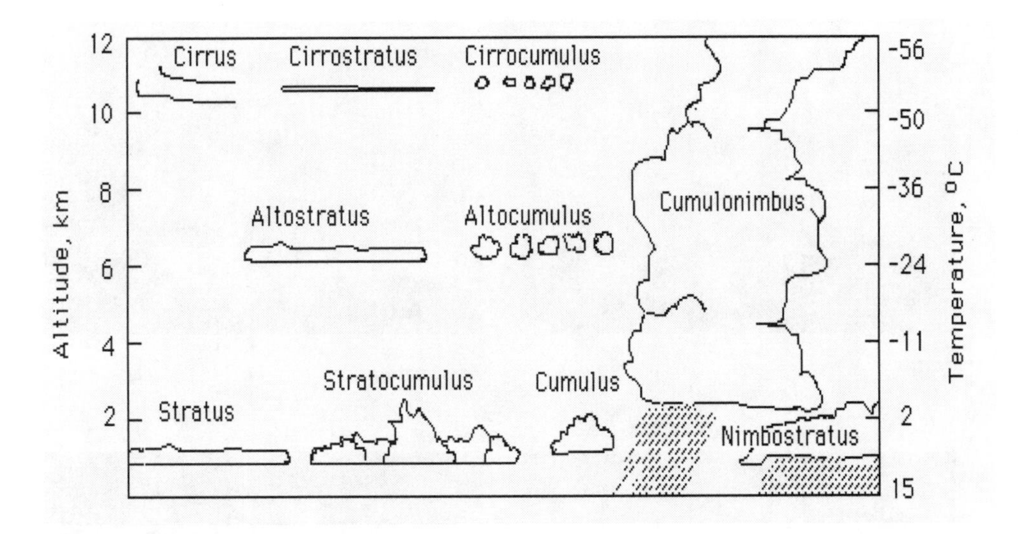

Cloud classification

The group of low clouds contains: *Stratus, Stratocumulus,* and *Nimbostratus.* The group of middle clouds contains: *Altostratus* and *Altocumulus.* The group of high clouds contains: *Cirrus, Cirrostratus,* and *Cirrocumulus.* The fourth group contains: *Cumulus* and *Cumulonimbus.* Low clouds are composed of water droplets. Within this group, stratus clouds appear as flat, whitish blankets. Stratocumulus clouds frequently cover the entire sky and look soft and gray. Nimbostratus clouds produce rain or snow, and are thick, dark and shapeless.

Middle clouds contain two principal genera. Altostratus clouds appear as a thick, gray or bluish veil. They softly diffuse sunlight or moonlight and sometimes produce a corona. Altocumulus clouds have the appearance of dense puffs.

High clouds are composed of ice crystals. They contain three principal genera. Cirrus clouds look like feathers, often with hooks. Cirrostratus clouds form fine, white veils. They often produce halos. Cirrocumulus clouds appear as small, white balls and wisps arranged in groups. They are popularly described as "mackerel scales".

Stratus

Stratocumulus

Altostratus

Altocumulus

Cirrostratus

Cirrocumulus

Cirrus

Nimbostratus

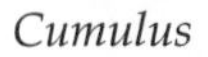

Cumulus

Cumulonimbus

6.22. MODEL OF A CUMULONIMBUS CLOUD

Materials: glass or jar, straw, water, milk, candle.

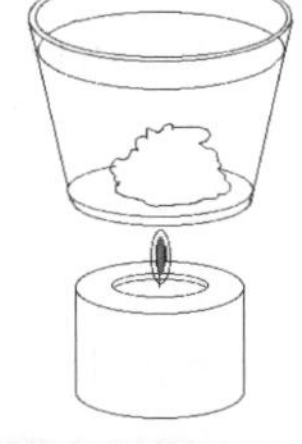

Procedure: Fill a glass with cold water. Let the water rest. Very slowly pour the cold milk through a straw to the bottom of the glass to make a layer about 2 cm deep. Place the glass over a burning candle. Observe a cumulus-like cloud forming in the glass.

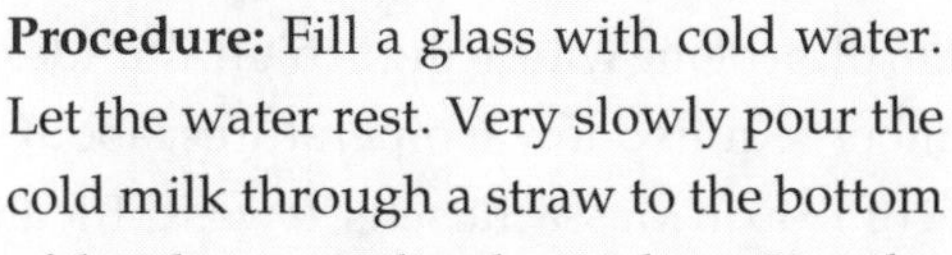

Comment: A similar experiment can be done by placing a jar, filled with hot, colored water, in a large tank with cold water. The jar should have a metal screw top with a few holes punched in it, to allow the hot water to rise.

Cumulus clouds have dense and well defined forms of individual domes or towers. They are associated with intense thermal updrafts. Cumulus clouds can grow into cumulonimbus, which are exceptionally dense and vertically developed. Their upper parts are likely to have the shape of an anvil. Cumulonimbus clouds produce thunderstorms with lightning, thunder and heavy precipitation.

The first meteorological cloud atlas was Hugo Hildebrandsson's *On the Classification of Clouds*, published in 1879. The first *International Cloud Atlas* was published in 1896. It included 28 color lithographs and was financed by Leon Teisserenc de Bort.

Contrails. Contrails are long streaks of cloud formed by high flying airplanes. Contrails usually appear about 8 - 15 km above the ground, where the air temperature is very low. Contrails are made of ice crystals. Most of the water used in their formation comes from an airplane engine exhaust.

RAINMAKING. On many occasions people have attempted to modify natural weather patterns by increasing rainfall or preventing the formation of large damaging hailstones. A variety of inventive, but non-scientific methods were used. They included burning fires, firing cannons, producing electric charges by kites, and ringing church bells.

In July 1946, two American physicists, Vincent J. Schaefer (1906-1994) and Irvin Langmuir (1881-1957) of the General Electric Company studied the problem of airplane wing icing at high altitudes. Experiments in a cloud chamber required low temperatures, about -23°C. During the hot summer of 1946 in Massachusetts, it was difficult to keep the chamber cold. In order to cool it, Schaefer used dry ice (frozen carbon dioxide). As a result, he accidentally discovered that tiny fragments of dry ice dropped into a cold chamber filled with a supercooled cloud caused the formation of several million ice crystals. On 13 November, 1946, the dry ice effect was tested in the real atmosphere. During this experiment, Schaefer dumped 3 kg of dry ice pellets from an airplane over a cloud in Massachusetts. The effect was the first man-made snowstorm in history.

Rainmaking experiments. In 1849, James P. Espy (1785-1860), the first official US meteorologist, attempted to generate artificial clouds and rain by setting fire to large tracts of forests. All his experiments failed.

6.23. SEEDING OF SUPERCOOLED CLOUDS

Materials: two metal cans, ice, salt, thermometer, insulating material, thick board, dry ice (warning: temperature of frozen CO_2 is about -70°C, consequently it should not be handled with bare hands), tongs.

Procedure: Construct a refrigerating chamber from two metal cans placed one inside the other. Fill the space between them with a 3:1 mixture of finely crushed ice and salt. The chamber should be encompassed by a thick insulating material. Until the temperature inside the inner can reaches -15°C, the top should be covered with a thick board. Breathing into the chamber produces a thick, gray, supercooled fog. If a small piece of dry ice is held over the cloud and scratched gently with a sharp point, a few specks of the material will fall into the cloud and show up as dense streaks resembling condensation trails. These contain thousands of tiny ice crystals produced by intense condensation followed by spontaneous freezing of droplets in a thin layer of air chilled by the particle of dry ice. The crystals will soon disperse through the chamber. The supercooled water droplets will evaporate and the vapor will be depositing on the ice crystals.

Comment: This experiment reproduces the process discovered by Schaefer and Langmuir.

Later, during the same year, minute crystals of silver iodide, were also found to act as efficient ice-forming nuclei, at temperatures below -5°C. On 21 December 1948, burning charcoal, impregnated with silver iodide was dropped from an aircraft into a supercooled (-10°C) stratus cloud , 300 m thick and covering approximately 13 km^2. The cloud was converted into ice crystals by less than 2 g of silver iodide.

Weather facts. If all the water vapor in the atmosphere could be artificially condensed, it would cover the entire Earth's surface with about 2 cm liquid water. Fortunately, this extreme event is not expected to happen, even though about 2000 thunderstorms occur every minute around the world. The maximum 24-hour precipitation (1870 mm of rain) was detected on Reunion Island (Indian Ocean) on 15 March 1952. One-year maximum precipitation (24,461 mm of rain) was recorded in 1861, in Cherrapunji (India). The record single storm snowfall, 480 cm, occurred on 13-19 February 1959, in Mt. Shasta, California. The wettest spot on Earth is Mount Waialeale, Hawaii, where there are 350 rainy days a year. The driest spot on Earth is the Atacama Desert in Chile, where in 1971 it rained for the first time after nearly 400 years of drought. Umbrellas, our personal safeguards against rain and sun, appeared in Europe in the 16th century. They were first used as sunshades. In the 18th century, umbrellas became fashionable as a protection against rain.

Everything should be made as simple as possible, but not simpler.
(A. Einstein)

CHAPTER SEVEN

FORCES

Ancient Ideas on Winds, Laws of Falling Bodies, Mathematical Developments. Mass, Energy, Beginnings of Hydromechanics, Newton's Laws of Motion, Atmospheric Forces, Gravitational Force, Pressure Gradient Force, Buoyant Force, Thermal Stability, Shear Forces, Centrifugal Force, Coriolis Force, Equations of Motion, Laminar and Turbulent Flows.

ANCIENT IDEAS ON WINDS. What causes winds? Early attempts to comprehend winds relied on various beliefs or speculations. For instance, some ancient Greeks believed that winds were drafts from the cave of the Greek god Aeolus. In Sicyon, an altar was raised to praise winds, and a year offering was made by night. In Delphi, a temple was specially dedicated to honor winds and the services they provided. Boreas, the stormy northern wind, represented as a gloomy-looking bearded figure, allegedly saved Athens. His friendly blast destroyed part of the Persian fleet of King Xerxes at Cape Sepias. In gratitude, an altar to Boreas was raised on the banks of the Ilissos.

The Greek philosopher Democritus (460-370 B.C.)

made use of his atomic theory by arguing that wind occurred whenever there were many particles (atoms) in a small space. When the space was large and the number of particles was small, the atmosphere was still and peaceful.

Aristotle (384-322 B.C.) argued that wind was air in motion resulting from the dry sighs of a breathing Earth. Regarding motion in general, Aristotle thought that it was produced by tendencies of each of the four elements, to move to its natural place. Water and earthly bodies tended to fall, while air and fire tended to rise. Any motion other than "natural" required the application of *force,* the pushing factor. Aristotle maintained that heavy objects fell faster than light ones. He also thought that a falling stone moved faster the closer it got to the ground, "*just as a horse moved faster when it approached its stable*".

For one thousand years, these opinions did not change significantly. In the Middle Ages, there were still those who believed that wind was caused by the flapping of angel wings.

Galileo's experiment

LAWS OF FALLING BODIES. According to Vincenzo Viviani, Galileo Galilei (1564-1642) carried two objects of the same shape but of different weight (a hundred-pound iron cannon ball and a one-pound ball) to the top of the 60-m high Leaning Tower of Pisa. In full view of the entire faculty and student body of the University of Pisa, he simultaneously dropped them to the ground in order to examine Aristotle's theory about falling bodies. The result of the experiment was that both balls reached the ground at the same time. Because there are no independent accounts of the event, historians doubt that this episode actually occurred. As the Italian proverb says: *Se non è vèro, è bon trovato,* "it might not be true but still it is a good story".

It is a fact that Galileo worked on the nature of vertical motion. In 1603, to make accurate measurements of quickly falling objects, Galileo devised a special apparatus. The construction consisted of a track, inclined at different angles, along which a bronze ball could roll down. The inclination served the purpose of slowing down the motion. For measurements of time, Galileo used a large bucket of water. A pipe of small diameter was soldered to the bottom of the vessel which dispensed a thin jet of water. The water was collected in a small beaker during the time of each descent. Its weight was assessed after each experiment on a very accurate balance and indicated the length of time interval.

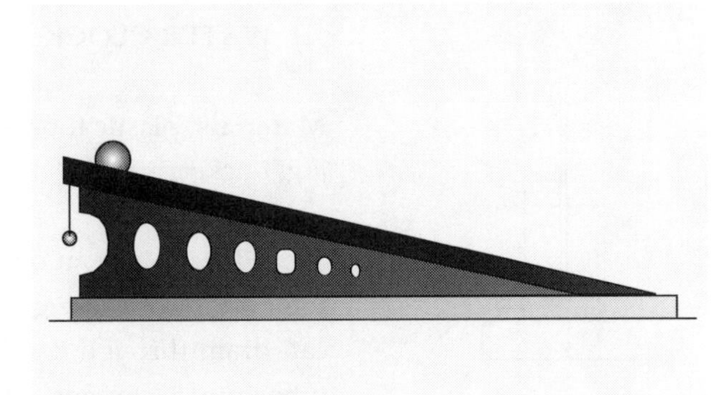

Galileo's device

After a series of experiments, Galileo found that the velocity of a body **v**, rolling from a resting position, was independent of weight but linearly increased with the time interval **t**: $\mathbf{v} = \mathbf{a}\,\mathbf{t}$, while the distance **s** increased as a square of the time interval: $\mathbf{s} = \mathbf{a}\,\mathbf{t}^2/2$, where **a** is a parameter (acceleration) dependent on a varying slope angle. Galileo correctly concluded that a free falling body would exhibit the same relation between distance, velocity and time. Presently, it is known that the value of the parameter **a** for a freely falling body is equal to 9.8 m/s^2, and is called gravitational acceleration. Gravitational acceleration is usually denoted by the letter **g**, so $\mathbf{g} = 9.8\ m/s^2$.

Galileo also observed that after the ball reached the bottom, it rolled along a smooth horizontal surface with a speed that was nearly uniform. As a result, he concluded that the velocity of a body did not change *"as long as external causes of acceleration or retardation are excluded"*. Consequently, he understood that acceleration or deceleration indicated the presence of force, and that a force had to be proportional to acceleration.

7.1. WATER CLOCK

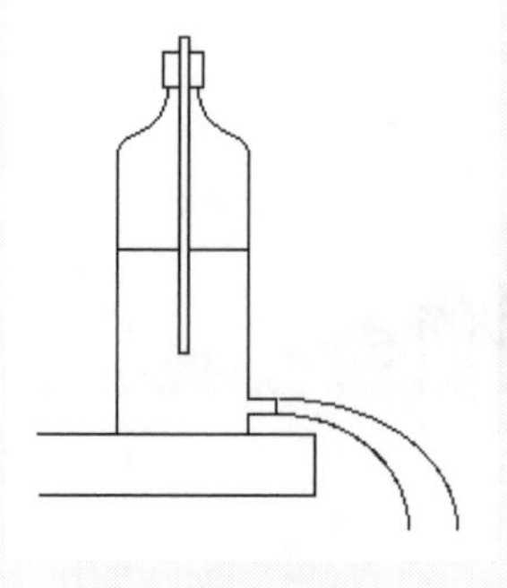

Materials: plastic tube, stopper, jug, bucket, water.

Procedure: Make an opening at the bottom of the jug so water can drain through it. Insert the tube into a stopper. Place the stopper into the jug. Notice that the velocity of the outflow does not depend on the level of water in the jug, but only on the distance between the bottom of the tube and the opening at the bottom of the jug. The tube can be moved up and down to obtain different flow rates. Collect the drained water in a bucket and weigh it. Time measured by this water clock can be evaluated as proportional to the mass of outflowed water (the proportionality parameter depends on the area of the opening at the bottom of the jug, the density of the liquid in the jug, and the distance between the lower opening and the bottom of the jug).

In the 1580's, Galileo made an observation of a swinging candelabrum (allegedly, while visiting a cathedral in Pisa). He timed the oscillations of the candelabrum by using his pulse as a watch. Later, this observation led the Dutch scientist, Christiaan Huygens (1629-1695), to conduct a study of the circular pendulum. Huygens' research resulted in the discovery of centrifugal force, as well as in the construction of the first *pendulum clock*. In 1673, Huygens published the results of his research on circular motion in the treatise *Horologium Oscillatorium*. He also noted that: *"if gravity did not act and air did not resist the motion of bodies, then any given body once set in motion would proceed along a straight line with uniform velocity"*.

MATHEMATICAL DEVELOPMENTS. Scientific progress in the seventeenth and eighteenth centuries was accompanied by substantial developments in mathematics. During this time, *logarithms* (from Greek words meaning "proportionate numbers") were invented by the Scottish mathematician John Napier (1550-1617). In 1614, Napier published his results in his famous book *Description of the Wonderful Rule of Logarithms.* Eight years later, in 1622 the English mathematician, William Oughtred invented logarithmic rulers. About 1644, Blaise Pascal built the adding machine (the Pascaline), the ancestor of all mechanical calculators.

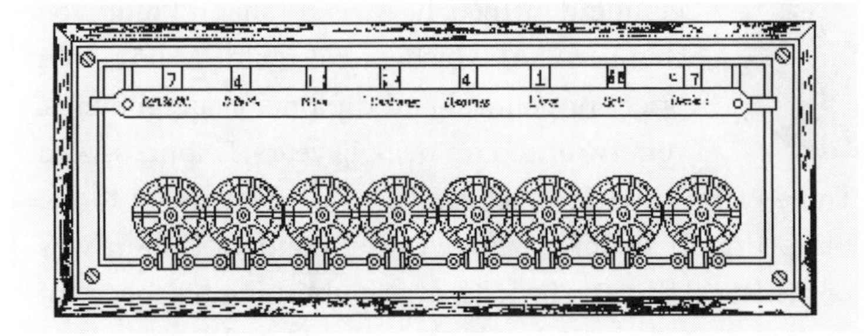

The Pascaline

In 1637, Rene´ Descartes (1596-1650) developed *analytical geometry* in his *Discourse on Method.* He defined a coordinate system with two *axes* X and Y, which were two perpendicular lines. The X-axis was horizontal, the Y-axis was vertical, and their intersection was called the *origin*. Any point P on the plane was represented by two *coordinates* (x, y). The coordinates (x, y) of all the points lying on a curve could be described by the *equation of the curve*. For instance, a straight line passing through the origin, and making a 45° angle with the X-axis is described by the equation: y = x.

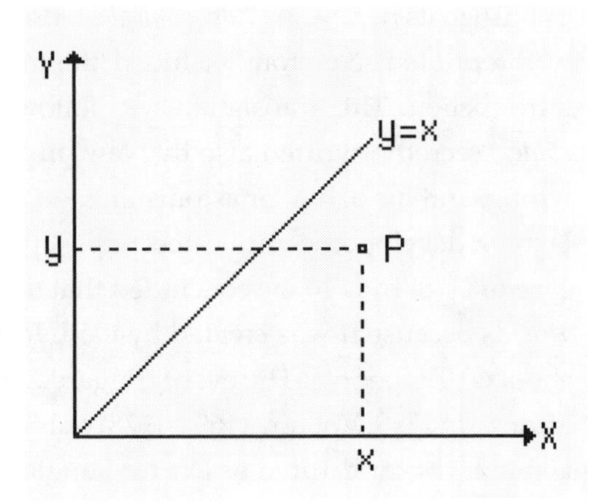

Cartesian coordinates

In 1694, the German mathematician, Gottfried Wilhelm Leibniz (1646-1716), introduced *function*, a concept of great significance in mathematics and science.

Gottfried Wilhelm von Leibniz (1646-1716)

Gottfried Wilhelm Leibniz, the German philosopher, historian, jurist, geologist, and mathematician, was born in Leipzig on 1 July 1646. He studied at Universities in Leipzig, Jena and Alfdorf. Beginning in 1666, he served Johann Philip von Schonborn, Archbishop Elector of Mainz, in a variety of political and diplomatic assignments. From 1676, and for over 40 years, Leibniz served Ernest Augustus, Duke of Brunswick-Luneburg (who later succeded to the throne of England as George I). In 1673, while visiting London, Leibniz was elected a member of the Royal Society. In 1684, after publishing his theory of calculus, Leibniz was accused of plagiarism by Isaac Newton. When Leibniz appealed to the Royal Society for a fair hearing, Newton, as President of the Society, appointed a committee of his own supporters and then wrote the committee's report, *Commercium Episolicum*. The report, of course, contained conclusions in Newton's favor. After it was printed, Newton anonymously published its review in *Philosophical Transactions*. When *Commercium Episolicum* was reprinted, Newton included this review in its Latin translation as an introduction. This translation was followed by an additional preface *To the reader*, secretly written also by Newton. Today, it appears that Newton was wrong, and his behavior astonishing.

Leibniz developed complicated philosophical theories to reconcile the existence of God. In 1710, he concluded that the Universe is the best of all possible worlds because it was created by God. In 1755, the Lisbon earthquake killed about 60,000 people. The event triggered attacks on Leibniz's optimistic point of view. In 1759, Voltaire (1694-1778) published the satiric tale *Candide*, in which Leibniz was caricatured as Doctor Pangloss.

Leibniz never married. He died neglected and forgotten in Hannover in 1716.

In 1665, Isaac Newton (1642-1727) invented *infinitesimal calculus* and *derivatives* (which he called *fluxions*) but kept the discovery to himself. Ten years later, in 1675, Leibniz derived the same method and introduced it to the world of mathematics.

Calculus. If y and x are two variables, and if a numerical value is assigned to x, then the defined value of y (dependent variable) is called a *function of x* (independent variable) and symbolized in writing as y = f(x). Calculation of derivatives is called *differentiation*, and the branch of calculus which deals with differentiation is called *differential calculus*. The derivative of function f with respect to x expresses how the function f(x) changes as x changes. Geometrically, the first derivative indicates a slope of a straight line which is tangent to the curve f(**x**) at a certain point **x**. The *velocity* can be understood as the first derivative of distance with respect to time.
There are several ways of denoting derivatives, invented by various mathematicians, as indicated below:

df/dt	$= D_x f$	$= f'$	$= \dot{f}$
Leibniz	Cauchy	Fermat, Lagrange	Newton

In Newton's notation, one dot was used for the first derivative, two dots for the second, and so on. Newton's notation was used exclusively in England until 1845. Any other notation was not acceptable, and it was not until 1845 in Cambridge, that one was allowed to use symbols other than Newton's.

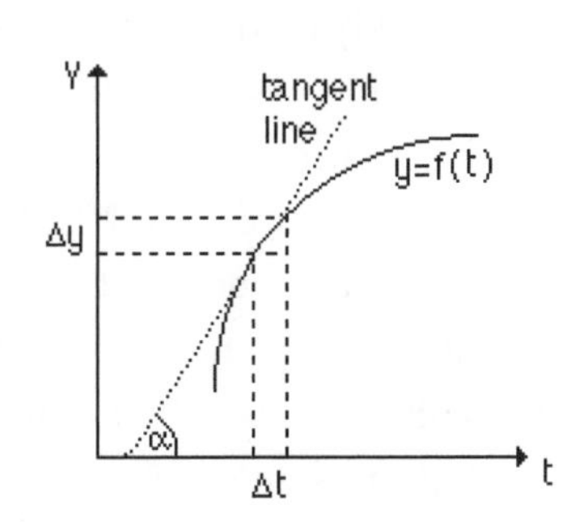

MASS. In 1671, the French astronomer Jean Richer (1620-1696) travelled to Cayenne in French Guiana, located near the Equator in South America. He took along a reliable pendulum clock. To his great surprise, in Cayenne, the clock ran slow, losing about two and a half minutes per day. To bring the clock into agreement with the astronomical time, Richer had to shorten the pendulum. When Richer returned to Paris, he observed that the clock gained back about two and one-half minutes a day.

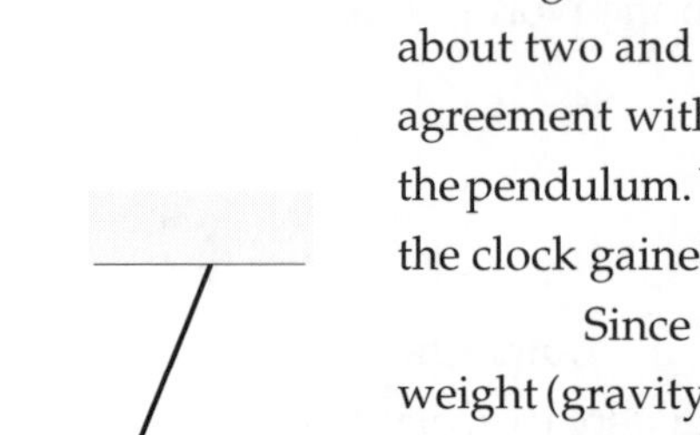

A pendulum

Since the motion of the pendulum was caused by its weight (gravity force), Richer's finding implied that the weight of the pendulum-bob had changed, even though it still looked like the same body. In spite of the change of the body's weight, something remained the same throughout the entire journey. To denote this constant property, Newton and Huygens applied the term *quantity of matter*, or *mass*. *Force* (e.g., weight) could be defined as a product of *mass* (mass of Richer's clock remained the same) and *acceleration* (its decrease caused the clock to run slow). From a dynamical point of view, *mass* is an important property because it is a measure of what might be called the body's "resistance" (called *inertia* - idleness) to being accelerated.

ENERGY. Gottfried Wilhelm Leibniz (1646-1716), while analyzing the impact of moving bodies, invented the term *vis viva* ("living force") for the quantity that is presently called *kinetic energy*. He defined it as a product of mass and height to which the body would rise as an effect of force. Using the laws derived by Galileo, he also showed that this height was proportional to the square of the velocity. A century later, Gaspard Coriolis (1792-1843) divided Leibniz's result by 2 and obtained the modern definition, $E_k = mv^2/2$, where m is the mass of the moving object, and v is its velocity. Christian Huygens (1629-1695) analyzed the motion of colliding bodies and concluded that the sum of *vis viva* of each ball was the same before and after the collision. It was a predecessor of the fundamental concept of the *conservation of energy*.

Lazare Leonard Carnot (1753-1823), the Minister of War during the French Revolution, made a contribution to science by defining the *latent vis viva*. Presently, the same quantity is called the *potential energy* ($E_p = \mathbf{mgh}$, where $\mathbf{m}$ is the mass, $\mathbf{g}$ is the gravity acceleration, and $\mathbf{h}$ is the height above a reference level). Carnot stated that any object raised to an elevated position gains the *latent vis viva,* because it can fall and change into the *vis viva* (kinetic energy).

In 1807, the Englishman, Thomas Young (1773-1829), a Professor of Natural Philosophy at the Royal Institution in London, proposed that the term *vis viva* be substituted with the word *energy*. Finally, William Rankine (1820-1872) introduced both terms, *kinetic and potential energy*.

BEGINNINGS OF HYDRODYNAMICS.
Evangelista Torricelli (1608-1647) is proclaimed by many as the founder of hydrodynamics. In a treatise *De Motu Gravium*, Torricelli dealt with the velocity of liquid spouting from a small opening in the bottom of a basin. He observed that if the liquid were made to spout upward, the jet reached a height slightly less than the liquid level in the basin. He deduced that if all resistance to the water motion were nil, the jet would reach a height of the liquid level in the basin. From this he derived that the velocity $\mathbf{v}$ of the jet at the point of outflow was equal to the velocity of a free falling droplet from the upper liquid level in the basin. Using laws derived by Galileo ($\mathbf{v} = \mathbf{gt}$, $\mathbf{h} = \mathbf{gt}^2/2$), Torricelli obtained: $\mathbf{v}^2 = \mathbf{g}^2\mathbf{t}^2 = \mathbf{g}^2 \times (2\mathbf{h}/\mathbf{g}) = 2\mathbf{gh}$, where $\mathbf{g}$ is the gravity acceleration, $\mathbf{t}$ is time, and $\mathbf{h}$ is the depth of water in the basin.

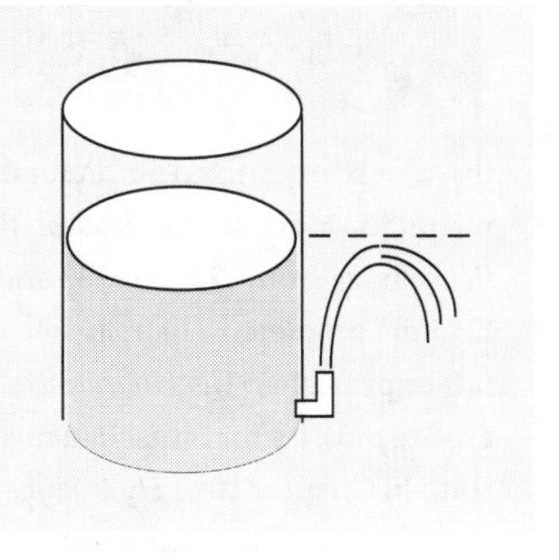

Torricelli's law

One of the predecessors of the science of fluids was Daniel Bernoulli (1700-1782). In 1738, Bernoulli published his *Hydrodynamica*. In this book, he formulated the principle which became known as *Bernoulli's Principle*. The original arguments regarding the principle were weak and were improved upon by his father, Johann I Bernoulli, in his *Hydraulica.*

Daniel Bernoulli (1700-1782)

Daniel Bernoulli was born on 8 February 1700 in Groningen in the Netherlands. He was the son of Johann I Bernoulli (1667-1748). Daniel was perhaps the most famous member of a family of well known scientists and mathematicians. Daniel was taught mathematics by his father and his older brother Nikolaus II. In 1724, Daniel published *Exercitationes Mathematicae,* which attracted much attention. In 1725, the newly organized St. Petersburg Academy of Sciences offered appointments to Daniel and Nikolaus II, which they accepted. Daniel worked in St. Petersburg from 1725 to 1733 and then moved to Basel in Switzerland. In 1724, Daniel completed *Hydrodynamica*, but it was not published until 1738. His father published his *Hydraulica* later, but predated it to 1732 in an attempt to ensure priority for himself (a rare example of competition between father and son). In Chapter 10 of *Hydrodynamica*, Daniel formulated the kinetic theory of gases. Daniel Bernoulli died in Basel, Switzerland on 17 March 1782.

The modern version of the Bernoulli Principle states that for frictionless motion which is time-independent, the total energy, comprising pressure, potential and kinetic energy, is constant. Bernoulli's Principle can also be formulated as follows: slow, horizontally moving air exerts more pressure than fast-moving air and vice versa (assuming frictionless and time independent flow). By this principle, Bernoulli was able to explain many phenomena of fluid motion, such as discussed in Experiments 7.2 - 7.5.

7.2. AERODYNAMIC PARADOX

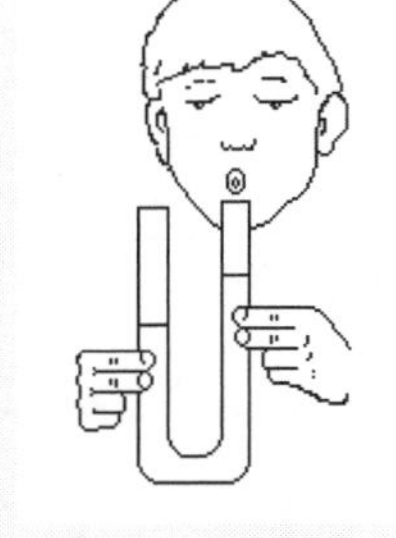

Materials: plastic tube, water.

Procedure: Shape the plastic tube in the form of a letter U. Fill it with water. Hold the tube so that both ends are on the same level. Put your lips next to one end of the tube and blow above it in a horizontal direction as hard as you can. Notice that the level of the water drops in the second end of the tube.

Explanation: The effect of the stream of air above the tube can be explained by Bernoulli's law. According to this law, the pressure in the air stream, which travels at a high velocity above the first end of the tube, is lower than above the other. The water in the tube moves toward the lower pressure, pushed by the atmospheric pressure.

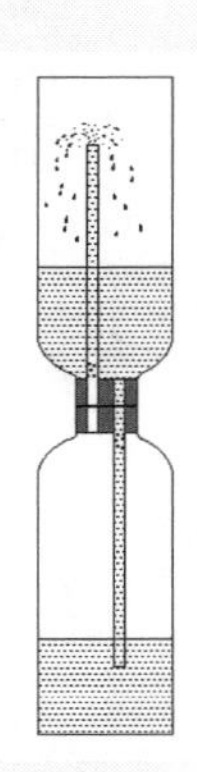

Comment: Based on the same principle, "Heron's fountain" (see page 70) can be built. It consists of two soda bottles with glued caps. Through the caps two holes are drilled to mount two plastic tubes. Each tube has a few little holes drilled near the mounted end. When water drains into the lower bottle it pushes the air into the upper one. Air rushes through the upper tube and picks some water with it through the small holes in the surface of the tube.

Amazing family. Daniel Bernoulli was a member of an amazing Swiss family of mathematicians, physicists and scientists. His father, Johann I, and his uncle, Jacob Bernoulli, set the foundation for the calculus of variations. Jacob is also remembered for his work on probability and calculus. Daniel's two brothers, a cousin, and a number of nephews were also mathematicians or scientists.

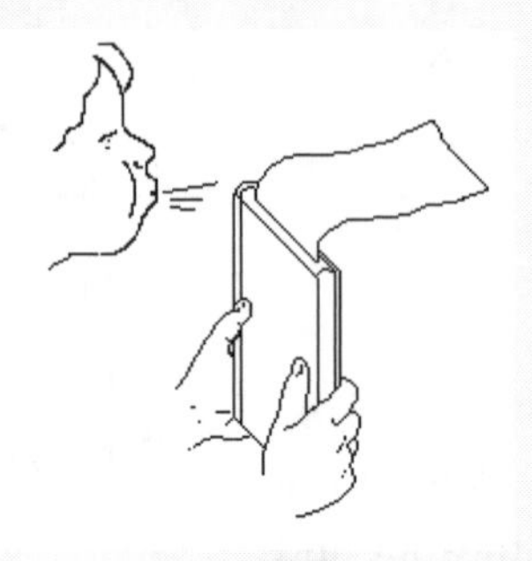

7.3. LIFTING FORCE (1)

Materials: sheet of paper, book.

Procedure: Insert one end of a sheet of paper inside a book. Hold the top of the book on the level of your mouth. Blow over the top of the sheet. The sheet of paper will rise.

Explanation: The fast-moving stream of air decreases the air pressure on the top surface of the paper, near the free end. The air pressure underneath the paper is greater and lifts the paper sheet.

Comment: A similar effect takes place when an airplane moves forward. A plane's wing is designed in such a manner that the air flows faster across its upper surface. This locally lowers the pressure above the wings and produces the lifting force that sustains an airplane in the air.
An analogous explanation applies to a sailboat when it moves in the direction nearly from which the wind is blowing. A locally lowered pressure on sail surfaces produces the horizontal force that pushes a sailboat.

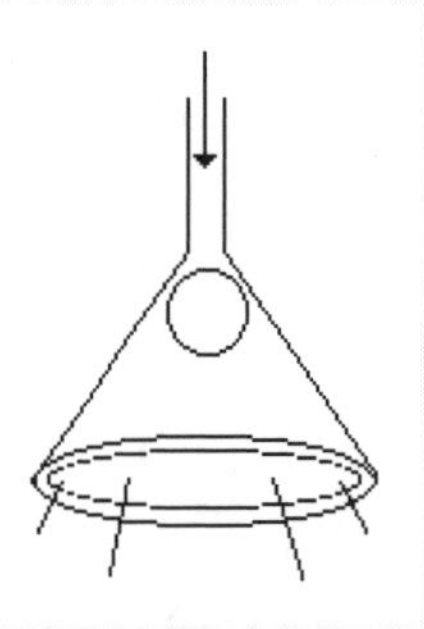

7.4. LIFTING FORCE (2)

Materials: ping-pong ball, funnel, air-pump (or fast stream of water).

Procedure: Pump air through the stem of the funnel. Place a ping-pong ball inside the funnel. The ball will not be blown away, but will be pushed into the funnel.

Explanation: According to Bernoulli's law, an air stream traveling at high velocity creates a region of low pressure in the funnel. The atmospheric pressure is higher outside and pushes the ping-pong ball toward the funnel.

7.5. LIFTING FORCE (3)

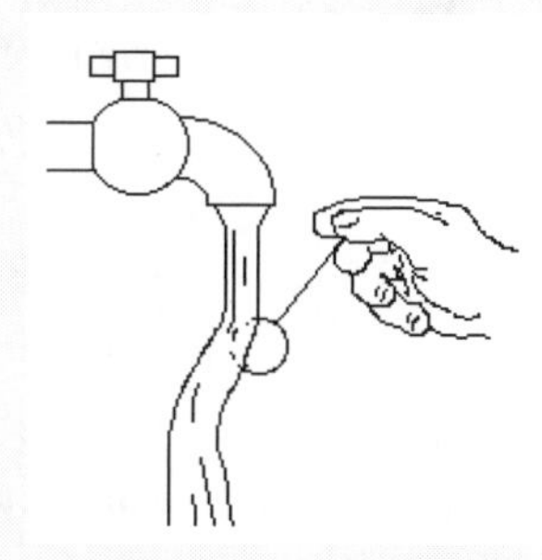

Materials: ping-pong ball, piece of string, glue, faucet.

Procedure: Glue a ping-pong ball to a piece of string. Place the ping-pong ball under a stream of water from a faucet. Observe that the ball is attracted to the stream of water.

Explanation: According to Bernoulli's law, the pressure in the water stream, traveling at a high velocity around the ping-pong ball, is lower than in the air. As a result, the atmospheric pressure pushes the ping-pong ball toward the water stream. Because of the resistance of the descending water, the ball cannot move all the way into the center of the stream.

NEWTON'S LAWS OF MOTION. In 1687, Isaac Newton published a treatise, *Philosophiae Naturalis Principia Mathematica*, generally referred to as *The Principia*. In this treatise, he formulated three laws of motion. According to the first law: "*Every material body persists in its state of rest or uniform motion in a straight line unless compelled by external force to change that state*". This means that acceleration is produced by force only, and also that force is the only factor which can produce acceleration. The first law implies that any body exerts a resisting force in opposition to any attempt to change its velocity. This property was earlier defined by Johann Kepler (1571-1630) as *inertia*. Hence, Newton's first law is sometimes called the *inertia law*.

Newton formulated the second law of motion as follows: "*Rate of change of motion* (acceleration) *is proportional to the motive force applied and takes place in the direction in which the force acts*". Newton's second law provides a method of measuring force as a product of mass and acceleration: **F** = **m a**, where **m** is the mass and **a** is the acceleration.

The third law of motion is as follows: "*The mutual action of two bodies is always equal and acts in an opposite sense*". A brief name for this law is the *reaction law*. When two bodies collide, they exert equal and opposite forces on each other and each changes the velocity of the other. The third law of motion implies the law of *conservation of momentum* (from a Latin word meaning "movement"). Newton defined *momentum* as *a product arising from the velocity and quantity of matter* (mass) *conjointly*" .

The Principia united terrestial and celestial physics and was considered the greatest in scientific achievements. Newton became a hero to the public. The British poet, Alexander Pope (1822-1744), wrote enthusiastically:

"Nature and Nature's laws lay hid in night:
God said, Let Newton be! And all was light."

ATMOSPHERIC FORCES. To understand complex atmospheric motions, imagine that air consists of many small parcels (for example small cubes). Each air *parcel* obeys Newton's laws of motion. Consequently, an air parcel of mass **m** has an acceleration **a** as a result of the action of various forces acting on it: $\mathbf{m}\,\mathbf{a} = \mathbf{F}_1 + \mathbf{F}_2 + \mathbf{F}_3 + ...$ Forces $\mathbf{F}_1$, $\mathbf{F}_2$, $\mathbf{F}_3$ ... may be classified into two categories, *body forces* and *surface forces*. Body (*gravitational*) forces act at a distance on the bulk of the air parcel (hence the word "body"). The surface forces (or stress forces) act by direct contact and are exerted directly on the surface of the air parcel. The surface forces can be divided into *normal stresses* and tangential *(shear) stresses*.

When talking about forces, it is necessary to introduce the concept of vectors. *Vectors* are quantities which are characterized by two, three, or more components. Quantities which are characterized by only one component are called *scalars*. Air density, pressure and temperature are scalars. Force, velocity and acceleration are vectors. In meteorology, vectors have three components to describe motion in three *independent* directions. Three components are necessary because space is three dimensional and any object in it is characterized by three sizes: length, width and height.

Vectors were introduced by Simon Stevin of Bruges (1548-1620), who was often referred to as Stevinus. In 1586, Stevinus published a book on statics and hydrostatics in which he used the principle of *vectors*, thereby giving a new impetus to science. Following Stevinus, Newton in his *Principia* gave the following corollary (see the figure above): "*Two forces conjoined will describe the diagonal of a parallelogram in the same time as it would describe the sides by those forces apart*". If the length of the first force is F_1, and the length of the second force is F_2, then the length F of the resulting vector is such that: $F^2 = F_1^2 + F_2^2 - 2 F_1 F_2 \cos \phi$, where ϕ is the angle between both vectors. When $\phi = 90^\circ$, then $\cos \phi = 0$, and the result is described by the Pythagorean theorem: $F^2 = F_1^2 + F_2^2$. Consequently, summation of forces is a vector summation, not an arithmetic summation.

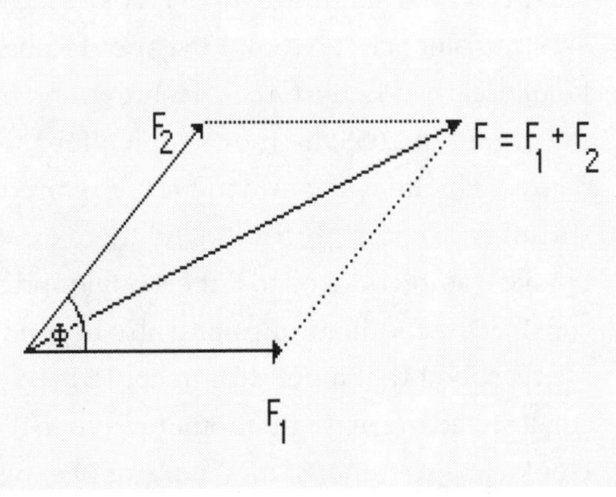

Sum of vectors

The following few sections contain a brief discussion of the major forces acting on the atmospheric air.

GRAVITATIONAL FORCE. Gravity is the result of an "attraction" between the air parcel and the Earth. The problem of gravity was discussed by Robert Hooke in 1682. In his lecture, *The Nature of Comets,* delivered before the Royal Society five years before the publication of Newton's *Principia,* he noted: "*By gravity, I understand such power as causes bodies to be moved one towards another till they are united*".

Robert Hooke (1635-1703)

Robert Hooke was born in Freshwater, on the Isle of Wight in England, the son of a minister. He was a sickly child and his parents did not have much hope for his survival. After the death of Hooke's father, Richard Busby, the master of the Westminster School, took 13 year old Robert into his home and educated him. Hooke learned Latin, Greek, Hebrew, and mathematics. He also learned to play the organ. In 1653, he moved to Oxford where he obtained a Master of Arts degree. Hooke was permanently impaired, his figure was crooked, his limbs shrunken. That is probably why there exists no original paintings of him. Hooke was introduced to Robert Boyle and became his assistant. In 1658, Hooke constructed the air-pump, and also began working with spring chronometers. He reasoned that a pendulum could be replaced by a spring. In 1658, Hooke constructed the plate anemometer. When the Royal Society was formed in 1662, Hooke was nominated as a "curator of experiments", and was asked to furnish three or four experiments at each weekly meeting. Hooke was able to accomplish this task, which provided an important intellectual stimulus for the Society. In 1665, Hooke published (and also illustrated) his *Micrographia* devoted to microscopical observations. In 1674, when Christian Huygens constructed a spring watch, Hooke cried foul and accused Henry Oldenburg, the secretary of the Royal Society, of being a "trafficker in intelligence" for revealing his secrets. In 1687, his niece Grace, first his ward and later his prolonged mistress, died. He never fully recovered from this loss, and even his interest in science ceased. He died on 3 March 1703, in the same room he had inhabited for nearly forty years, at Gresham College in London.

In *his Principia* (1687), Newton formulated the law of gravitation as follows: "*There is a gravitational attraction between every two bodies in the universe, and this force is proportional to the quantity of matter (mass) of each, and varies reciprocally as the square of its distance*". Mathematically, this law can be expressed as: $\mathbf{F = G\,m\,M/r^2}$, where m and M are masses of two bodies, r is the distance between them, and G is a constant.

Newton's apple. The idea of gravitation came allegedly to Newton in his garden, as he *"sat in a contemplative mood and was occasioned by the fall of an apple"*. According to Newtons' nineteenth century biographer, David Brewster, this story was passed on by Catherine Barton, Newton's niece, to Voltaire, who circulated it. Brewster also claimed that he saw Newton's apple tree before it was cut down in 1820. There exists, however, a preserved correspondence between Hooke and Newton, dated 1679, in which Hooke states his conviction that gravity decreases proportionally to the inverse of the square of the distance. After the publication of Newton's *Principia* in 1687, Hooke claimed that Newton had stolen the idea of the inverse square relation from him. According to historians, Hooke did not discover the law of universal gravitation, but did set Newton on the correct track.

Ocean tides. Based on his law of gravitation, Newton offered a correct explanation that ocean tides were caused by the gravitational attraction among the Earth, Moon and Sun. Since the force of gravity decreases with the distance from the Sun (or the Moon), the force acting on the ocean water on the closer side of the Earth is larger than the force acting on the solid body of the Earth. Similarly, the force of gravity acting on the ocean water on the farther side is smaller than that acting on the solid body of the Earth. As a result, the water surface on the closer side has the tendency to rise higher over the ocean bottom, while the far side of the Earth is accelerated away from the water. Both effects result in the formation of two water bulges, which in conjunction with the Earth's rotation around its axis, are observed as two tidal waves running around the Earth within a period of approximately 24 hours.

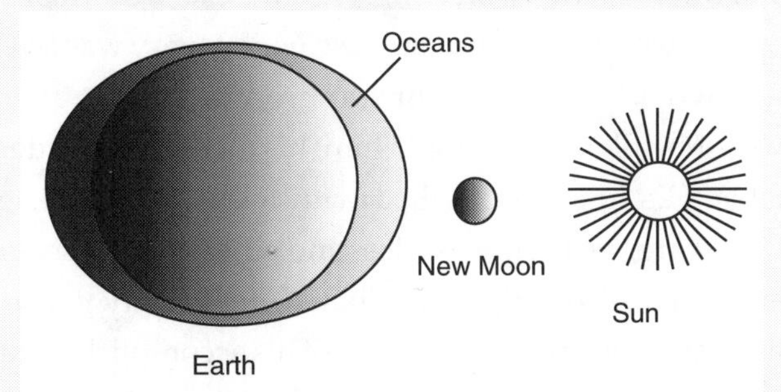

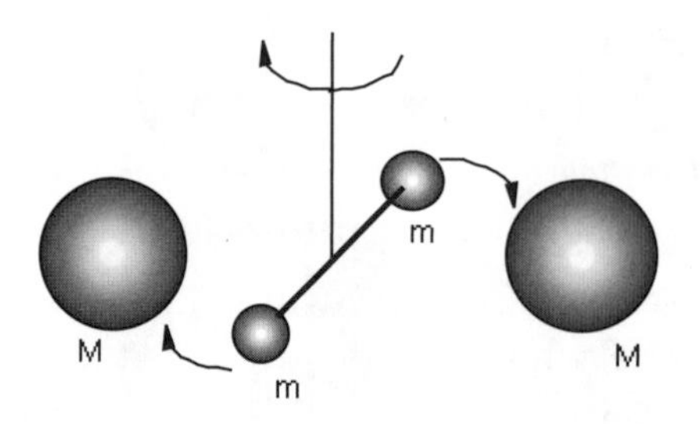

Cavendish's experiment

In 1798, Henry Cavendish (1731-1810) set up a very delicate torsion balance and directly measured the gravitational force of attraction between two pairs of lead balls of known masses and distances apart. As shown in the figure, the balls of the first pair (m) were placed on either end of a suspended beam. These balls were attracted by a pair of stationary lead balls (M). Cavendish calculated the force from the amount of twist from a wire and found that $G = 6.48 \times 10^{-11}$ $N\ kg^{-2}m^{-2}$ (where N is a symbol of a unit of a force, newton, defined on page 62). The presently accepted value of G is $6.67 \times 10^{-11}\ N\ kg^{-2}m^{-2}$.

Henry Cavendish (1731-1810)

Henry Cavendish was born in Nice, France on 10 October 1731. His father, Lord Charles Cavendish, was the son of the second Duke of Devonshire, and his mother was the daughter of the Duke of Kent. Cavendish senior was a distinguished experimenter and prominent figure in the counsels of the Royal Society. He frequently took his son Henry to meetings of the Royal Society and to dinners at the Royal Society Club. Henry was elected to membership of these organizations in 1760, and rarely missed a meeting. His interests were in pure mathematics, mechanics, optics, geology and industrial science. Henry was a very shy and reserved person who would dash away when approached by a stranger. Even his female servants were not allowed to talk to him; his meals were ordered by notes left on the hall table. His house was filled with telescopes, packages of chemicals, chemical glassware, measuring devices and other kinds of scientific apparatus. Cavendish isolated hydrogen and evaluated the gravity constant. Nevertheless, he published only a few papers. Most of his accomplishments were unknown, until one hundred years after his death, when James Maxwell published his notes.

Using his result, Cavendish was able to evaluate the mass **M** of the Earth. He used Newton's second law: **F = mg**, and Newton's law of gravitation: **F = G mM/r²**. Since the force described by both laws is the same, it can be obtained that **G mM/r² = mg**, or **M = g r² / G** = 9.8 x (6,370,000)² /(6.67 x 10^{-11}) kg = 5.84 x 10^{24} kg. On page 56, the mass of the atmosphere was evaluated at 5.25 x 10^{18} kg, indicating that the Earth is a million times heavier than the atmosphere.

The alternative law of gravitation. Some (less than serious) physicists argue that Newton's law of gravitation should read: *"Dropped objects fall in such a way that the resulting damage is the largest"*. As proof, it is said that 89% of open face sandwiches, that are accidentally dropped, land on their buttered sides.

On the other hand, from the equation **GmM/r² = mg**, it can be obtained that **g = GM/r²**. This expression indicates that gravity acceleration is not constant but depends on the distance **r** from the center of the Earth. The Earth is not a perfect sphere but an oblate spheroid; it flattens at the poles and bulges near the equator. The equatorial radius is about 208 km greater than its polar radius. As a result, gravity acceleration **g** on the Earth's surface depends on latitude. This explains the earlier noted discrepancies of the pendulum clock readings which were observed by the French astronomer Jean Richer in Cayenne and Paris (see page 192).

Effect of pressure force

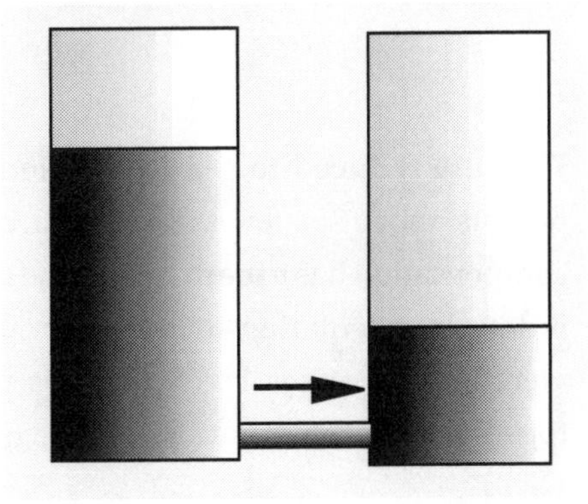

PRESSURE GRADIENT FORCE. The pressure gradient force acts from higher to lower pressure and is proportional to the *pressure gradient*, i.e. the rate at which pressure changes in space. For instance, if in two containers of water connected by a pipe, there is a difference in water levels, the pressure difference will create a force which will move water from the container with a higher level to the other one, and the motion will continue, until both levels of water are on the same plane.

As proved in 1648 by Blaise Pascal (1623-1662), the atmospheric pressure decreases with height. As a result, one might expect that the pressure force should be able to move the atmospheric air out to space. But this does not occur. The explanation of this fact was first given by Pierre Simon Laplace (1749-1827). In his work entitled *Celestial Mechanics*, published in 1823, Laplace noted that the vertical pressure gradient force is approximately balanced by the gravity force. This balance is expressed by the so called *hydrostatic equation*, which relates the air pressure and altitude at two levels. The hydrostatic equation can be written in the following form: $(p_2 - p_1) / (z_2 - z_1) = -\rho g$, where z is height, ρ is the air density, p is the air pressure, g is the gravity acceleration, and the subscripts refer to two levels in the atmosphere.

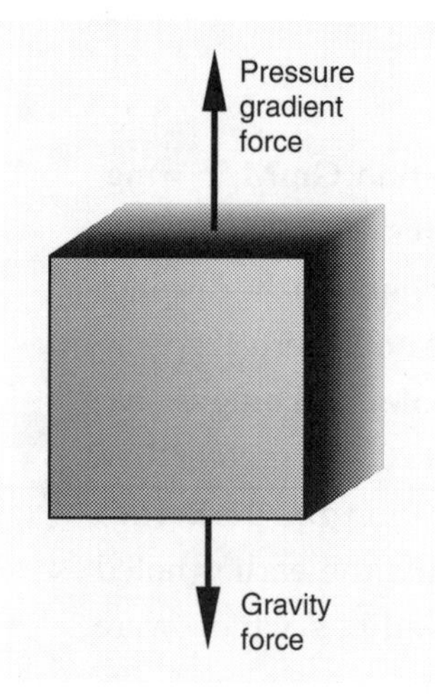

Hydrostatic balance

Since the air density ρ depends on temperature, thus the vertical pressure gradient is also dependent on temperature, and varies with height. Near the Earth's surface the vertical pressure gradient is about 10 millibars per every 100 meters.

Pressure reduced to sea level. Meteorological stations are located at various heights above sea level. To compare pressure measurements at these stations, compensation has to be made for the elevation of each station. This compensation is done by calculating an imaginary pressure which air would have if the station were located at sea level. For this purpose the hydrostatic equation is used together with the equation of state (see page 79).

7.6. ANALOGY OF HYDROSTATIC BALANCE

Materials: hair dryer, inflated rubber balloon.

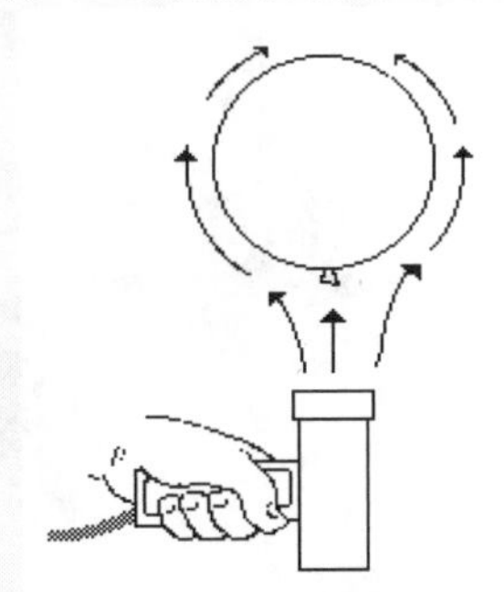

Procedure: Place an inflated rubber balloon in the stream of air generated by a hair dryer. The balloon floats in the stream of the air.

Explanation. The balloon floats because its gravity force is balanced by the pressure gradient force generated by the stream of air. The balloon is located at the distance from the hair dryer where the stream velocity is low enough to allow the equilibrium.

Comments: (i) The balance of forces acting on the balloon is analogous to the *hydrostatic* balance of the pressure and gravity forces in the atmosphere.
(ii) Notice that the balloon floats in the stream also when the direction of the stream is not vertical (due to Bernoulli's effect).
(iii) A similar experiment was performed by Heron (about 60 B.C.), who was the first to show that a small ball can float in a stream of vapor ejected from a container of boiling water.

BUOYANCY FORCE. The presence of the buoyancy force was discovered by one of the greatest philosophers of the ancient world, Archimedes of Syracuse (287-212 B.C.), student of Euclid at the University of Alexandria. Archimedes described his discoveries in his famous treatise "On floating bodies". One of the theorems in his treatise reads: "*if any body lighter than a fluid is forcibly immersed into it, the body will be driven upwards by a force equal to the difference between its weight and the weight of the fluid displaced by the immersed body*".

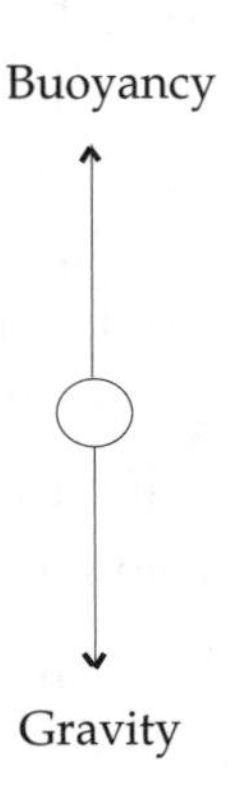

A golden crown. Archimedes was the nephew of King Heron of Syracuse, Sicily. Legend says that King Heron ordered his jeweler to make him a new crown of a weighted quantity of gold. When the crown was made and delivered, the king demanded that its weight be examined. The crown was tested and it weighed exactly the same as the original piece of gold. Nevertheless, the King was haunted by a suspicion that the rascally goldsmith had mixed silver with the gold to produce the crown. However, he did not know how to prove it. Consequently, Archimedes was asked to solve the puzzle. Archimedes allegedly found the solution to the problem during his visit to the public baths. When a servant filled the bathtub brimful of water and Archimedes got in, water flowed over the brim. Suddenly, the idea struck him that the weight he had lost upon submersion was equal to that of the water which brimmed over the tub. In his joy of making this discovery, he jumped out of the tub, forgot to put on his clothes, and ran through the city exclaiming *eureka, eureka,* which means "I found it". After performing a series of experiments, he solved the puzzle and, in fact, proved that the crown was not made of pure gold.

Let us try to follow Archimedes' way of thinking. Suppose the crown weighed 789 g in air and 739 g in water. The density of gold is 19.3 g/cm^3, of silver 10.5 g/cm^3, while of water 1 g/cm^3. Consequently, Archimedes could have obtained:

- loss of weight in water = 789 g - 739 g =50 g,
- weight of water displaced by the crown = 50 g, so volume of crown = 50 g /(1 g/cm^3) = 50 cm^3,
- crown's weight if made of pure gold = 50 cm^3 x 19.3 g/cm^3 = 965 g,
- shortage in weight = 965 g - 789 g = 176 g,
- substituting 1 cm^3 of silver for gold reduces the weight by 19.3g -10.5 g = 8.8 g,
- volume of silver = 176 g /8.8 (g /cm^3) = 20 cm^3,
- volume of gold = 50 cm^3 - 20 cm^3 = 30 cm^3.

Therefore, the crown was made of 30 cm^3/50 cm^3 = 60 % of gold, and 40 % of silver. The goldsmith had stolen 789 g x 40% = 315.6 g of gold!

7.7. BUOYANCY FORCE

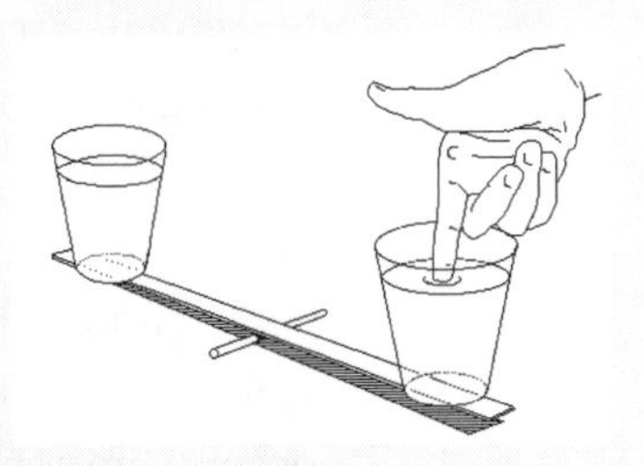

Materials: two glasses, long ruler, pencil.

Procedure: Place a glass on each end of a ruler, with a pencil beneath, to form a crude balance scale. Adjust the pencil until the scale is almost balanced. Now plunge your finger into the water in one of the glasses, without touching its walls. The buoyancy force exerted by the finger will immediately tip the "scale".

7.8. ZERO BUOYANCY

Materials: jar, piece of wax.

Procedure: Place a piece of wax in water. Note that it floats. However, when it is held for a moment at the bottom of the jar, it remains at the bottom.

Explanation: When the piece of wax is at the bottom, water is unable to push on it from below and lift it.

The buoyancy force is related to differences of density. In the atmosphere, differences in air density can occur in various ways, for instance, through changes in temperature, or changes in content of water vapor. Because warm air is lighter than cold air and humid air is lighter than dry air, density changes can lead to vertical motions called *convection*.

THERMAL STABILITY. When an unsaturated parcel of air moves upwards, it expands due to the decrease of the atmospheric pressure with height. The process of expansion decreases the internal energy of parcel molecules, and causes the parcel to cool. This cooling takes place at the constant rate of about 1°C per 100 m, which is called the *dry adiabatic lapse rate.* Analogously, the sinking air warms with the same *dry adiabatic lapse rate.*

As explained on page 100, the environmental temperature lapse rate varies. Therefore, sometimes the rising air might be cooler, and therefore heavier, than the surrounding air. Such air is stopped in its upward motion, and then forced to sink. When this occurs, the conditions are called *stable.*

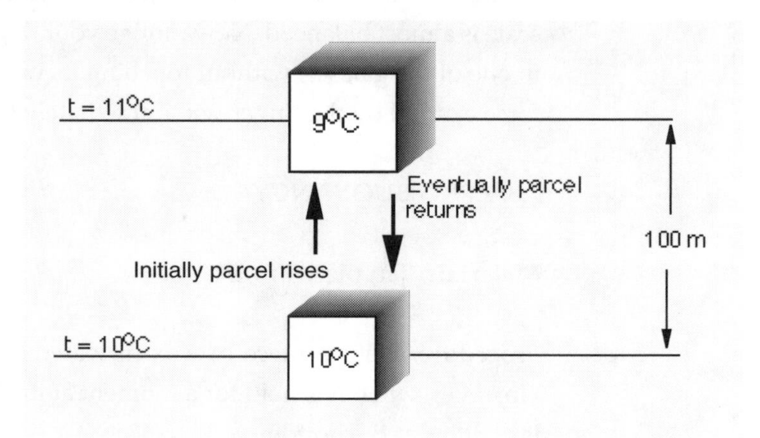

Buoyancy in the stable atmosphere

Such a case is considered in the figure above. The temperature in the atmosphere increases with height from t_1 = 10°C to t_2 = 11°C. A parcel, which initially has a temperature of t_{p1} = 10°C, is moved upwards. As a result, it expands and cools adiabatically to t_{p2} = 9°C. Being cooler than the ambient air, the parcel returns to the initial level. If the temperature in the figure decreased with height from t_1 = 10°C to t_2 = 8°C instead, the rising air at the upper level would be warmer (t_{p2} = 9°C), and therefore lighter, than the surrounding air (t_2 = 8°C). As a result, the parcel would continue rising. Consequently, this state is called *unstable.* In the *neutral* case, the rising particle has the same temperature along its path as the surrounding air ($t_1 = t_{p1}$, $t_2 = t_{p2}$, the temperature lapse rate is 1°C/100 m).

When the air moves over a hill (figure on the next page), the thermal stratification determines the patterns of such a flow. Imagine that an air parcel is moving up a hill. On the top, the parcel is cooler (adiabatic cooling) than at the bottom. In the stable atmosphere, the parcel is cooler than the ambient air and sinks along the hill. In the unstable atmosphere, the parcel is warmer and continues rising.

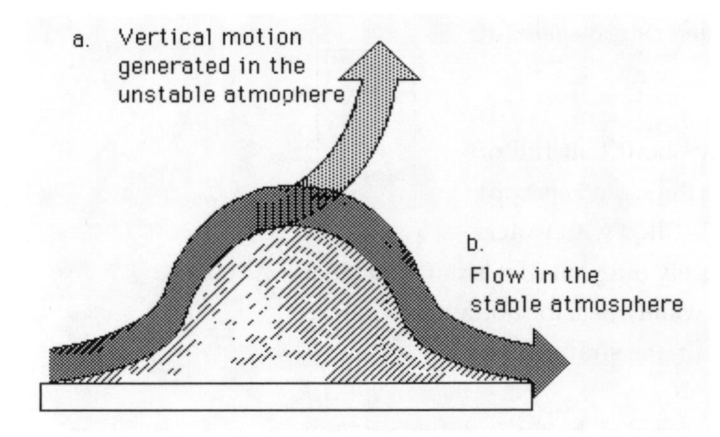

Flow in: (a) the unstable, (b) the stable atmosphere

7.9. BUOYANT STABILITY (1)

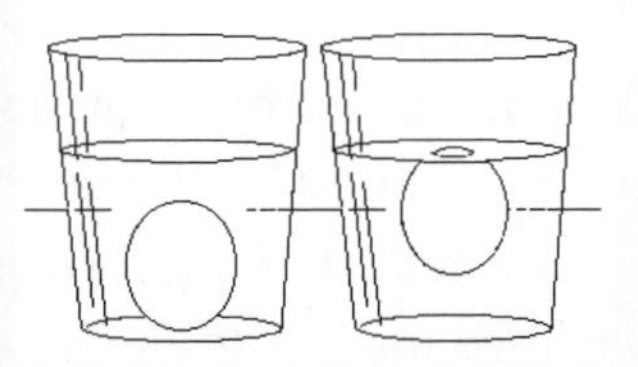

Materials: two glasses, water, salt, two fresh eggs, ice cubes, spoon.

Procedure: Fill two glasses about half full of water. Add salt to the water in one glass, a tablespoon at a time, and stir it. Continue adding and stirring until no more salt can dissolve. Place an egg into each glass. The egg in the heavier, salted water, will float. The egg in the fresh water will sink. Put a large ice cube into each glass. Note that the ice cube in the salted water protrudes more from the water.

Comment: Note the analogy to the atmospheric thermal stability. The fresh water represents the warm and lighter air. The salted water represents the cool and heavier atmospheric air. The eggs represent a parcel of air. The phenomenon occurs also in the oceans. The differences in the saltiness (salinity) of sea water affect the stability of a water column.

7.10. BUOYANT STABILITY (2)

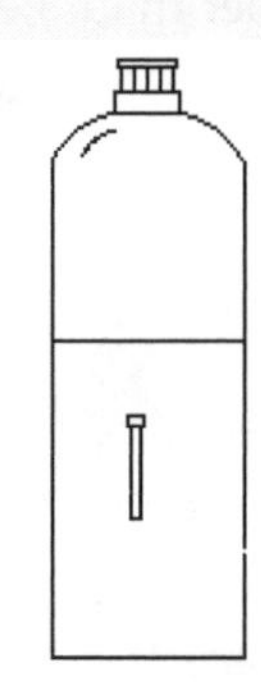

Materials: large plastic soda bottle with a sealing cap, two-centimeter straw sealed at one end.

Procedure: Fill a straw about half full of water. Let it float (with the sealed end up) in the plastic soda bottle filled with water. The straw should be barely buoyant (if it is not, you have to add or remove some water). Seal the bottle and squeeze the sides of it. As a result, the straw will sink.

Explanation: When the sealed bottle is squeezed, the pressure in it increases. This forces water to enter the straw. The buoyancy force decreases and the straw sinks.

Comments: (i) Note an analogy to atmospheric convective motions caused by thermal stability. (ii) The experiment was allegedly first performed by Descartes, and therefore is named the "Cartesian diver".

On a hot day individual "bubbles" of air begin rising after being thermally (heated from the surface) or dynamically (flow over obstacles) initiated. Shallow layers of unstable air produce small, puffy, cumulus clouds. Deep layers of unstable air produce large, towering cumulus clouds. Therefore, generally, cumulus clouds indicate unstable conditions.

The content of water vapor in the atmosphere has an important impact on the thermal stability. When the rising parcel of air reaches its dew point, vapor condenses and releases the latent heat of condensation. The released heat increases the air parcel's temperature. As a result, the adiabatic lapse rate decreases from 1°C per 100 m to about 0.6°C per 100

m. This new value is called *pseudo-adiabatic lapse rate* or *moist adiabatic lapse rate.*

The release of latent heat causes the rising parcel to intensify its vertical motion. This mechanism provides additional "fuel" for the formation of thunderstorms.

SHEAR FORCES. Molecules or parcels of gases or liquids moving with different velocities generate friction forces on the surfaces between them. In 1687, in his *Principia,* Newton proposed that "*the resistance arising from the want of lubricity in the parts of a fluid is proportional to velocity with which the parts of the fluid are separated from each other*". Today, we would formulate this as the stress τ, i.e., the shear force per unit area (Newton's "*resistance arising from the want of lubricity*") is proportional to the velocity change **U** along a distance **d**, $\tau = \nu \mathbf{U}/\mathbf{d}$ (see the figure below). The proportionality coefficient ν is called the *viscosity.* Viscosity is a property of a fluid (i.e., gas or liquid) and is independent of the flow geometry.

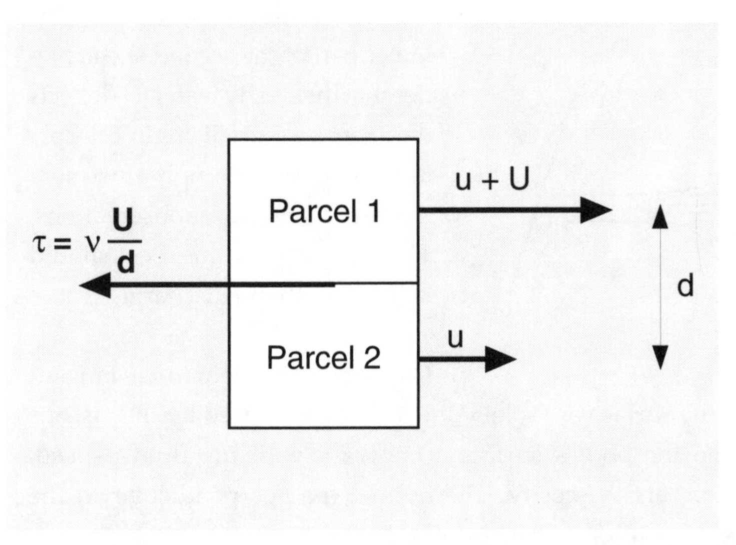

Shear force

7.11. SHEAR FORCES (1)

Materials: two thin, flat rectangular, 20 cm x 5 cm sheets of metal, two 0.1 m pieces of strong metal wire, rubber hose, water supply.

Procedure: Sharpen the long edges of the first metal sheet. Drill a hole close to the middle of the shorter edge of the sheet, just large enough to insert the wire. Insert the wire through the hole and use glue to mount it perpendicular to the faces of the sheet. Attach a piece of wire along one of the short edges of the second remaining metal sheet. Place the first sheet in a water jet in such a way that water flows on both sides of the sheet. Strengthen the jet until the sheet is visibly deflected. The sheet deflects more when speed of water increases.

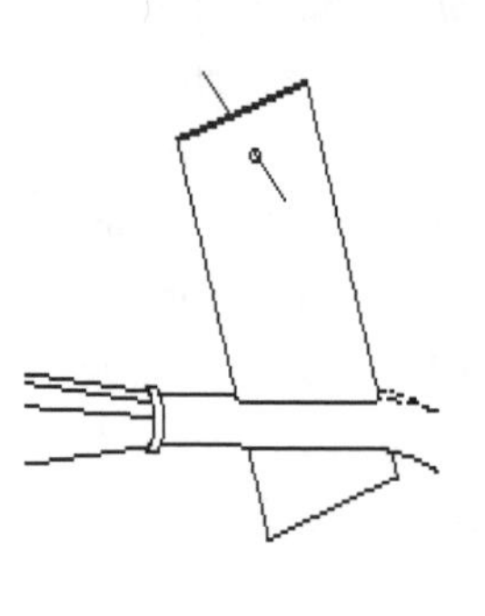

Next, direct a jet of water against the lower part of the second sheet. Notice that the steady water jet deflects the sheet by a small angle. Sharply increase the water supply a few more times, repeating these observations. The deflection of the sheet should increase with the discharge.

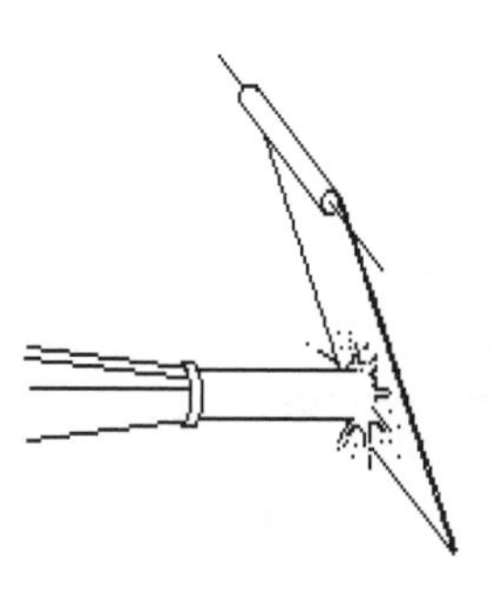

Comment: The shear forces tangential to the surface are solely due to friction exerted by the viscous fluid on the body's surface. They vary with the fluid's speed, direction, and viscosity. The normal (i.e., perpendicular to the surface) shear forces consist of two parts. The first part is proportional to pressure, and the second one depends on fluid viscosity and changes in speed perpendicular to the surface.

7.12. SHEAR FORCES (2)

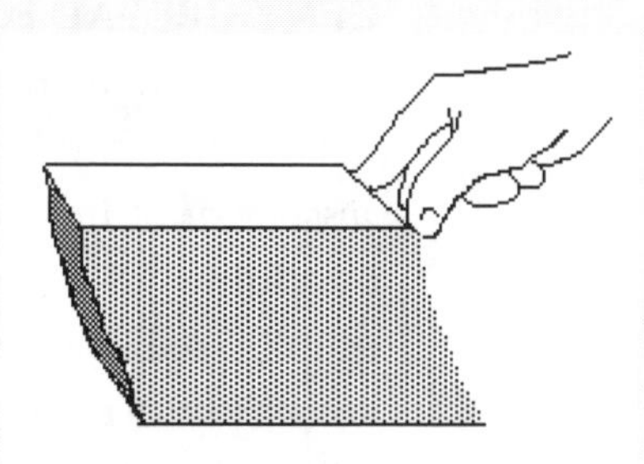

Materials: foam rubber.

Procedure: Place a block of the foam rubber on the table. Push it from a side in a direction parallel to the table. Observe the resulting deformation.

Explanation: If a sufficient tangential force is applied, a block of foam rubber can undergo large shearing deformation. The block resists against pushing and it returns to the initial shape when the tangential force is not applied. Similarly, fluids resist being changed in shape.

7.13. FLUID VISCOSITY

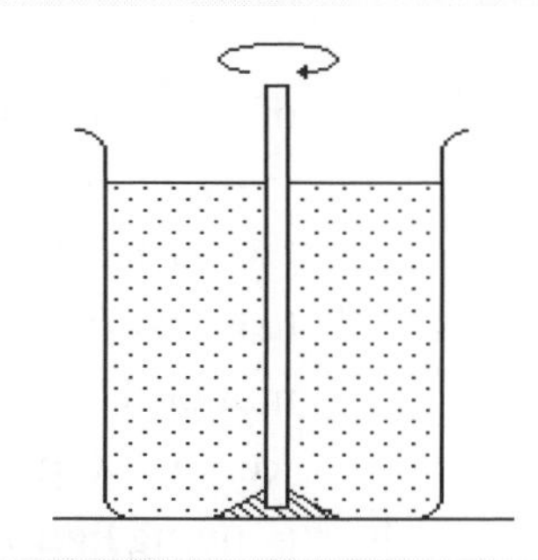

Materials: two half-liter beakers, 25 cm-glass rod, water, cooking oil , modeling clay.

Procedure: Mold a cup-shaped depression in the modeling clay with the end of the glass rod.

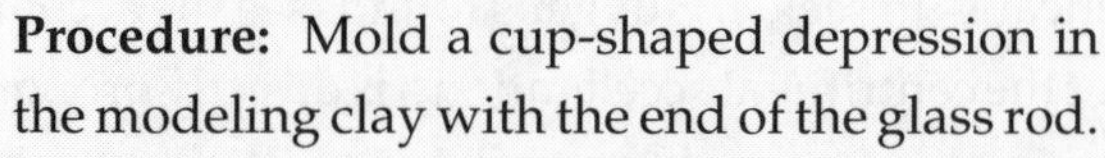

Place it in the center of the bottom surface of the first beaker. Add cooking oil until the beaker is about two-thirds full. Stir briskly so that air bubbles form (they will serve as flow markers). Now place the rod upright in the modeling clay cup. Begin slowly and steadily rotating the rod in one direction. Initially only bubbles near the rod are set in motion, but with time the movement will increasingly spread out. Eventually, the bubbles will travel over circular orbits. Repeat this procedure with water in the second beaker.

Explanation: Notice that some effort is required in order to stir each liquid. It is easier to stir water than oil. According to Newton's hypothesis on page 211, resistance force which arises in a liquid is proportional to the velocity gradient along the radius, but it also depends on a property (viscosity) which is specific to the liquid.

CENTRIPETAL FORCE. Velocities and accelerations in Newton's laws of motion are measured within a reference frame which does not accelerate. When a reference frame is accelerated in relation to *absolute space*, the form of Newton's laws of motion must be changed by adding new terms. For instance, when the Earth is the reference frame, the absolute velocity of the air is the vector sum of the relative velocity with respect to the Earth and the Earth's rotational velocity with respect to the absolute space. Acceleration in a rotating system differs from that in absolute space by two additional terms: the *centrifugal* and *Coriolis accelerations*.

As described in 1659 by Christiaan Huygens, the centrifugal force (in Latin *fugio* means "to flee") is an apparent force which acts in a rotating system, perpendicular to the axis of rotation. For example, when an object at the end of a string is swung around in a circle, the tightness of the string seems to indicate the presence of the centrifugal force. But one can also take a different point of view. In accordance with Newtonian mechanics, the swung object has the tendency to move straight forward. Thus, there is a real force (e.g., exerted by a string) that pulls the object towards the center of motion. Huygens called it the *centripetal force* (in Latin *centrum* means "center", and *peto* means "to seek"). Huygens found (see the accompanying figure) that the centrifugal acceleration a is directly proportional to the squared velocity v and inversely proportional to the radius **r**: $a = v^2/\mathbf{r}$. Huygens was able to calculate the centripetal acceleration at the Equator. For the radius of the Earth R = 6370 km, the time of one Earth's revolution T = 24 hours, one obtains: $v^2/R = (2\pi/T)^2 \times R = (2\pi/86{,}400)^2\, 6{,}370{,}000\ m/s^2 = 0.0337\ m/s^2$. The result indicates that the centripetal acceleration at the Equator is about 300 times smaller than gravity acceleration, which is $9.8\ m/s^2$.

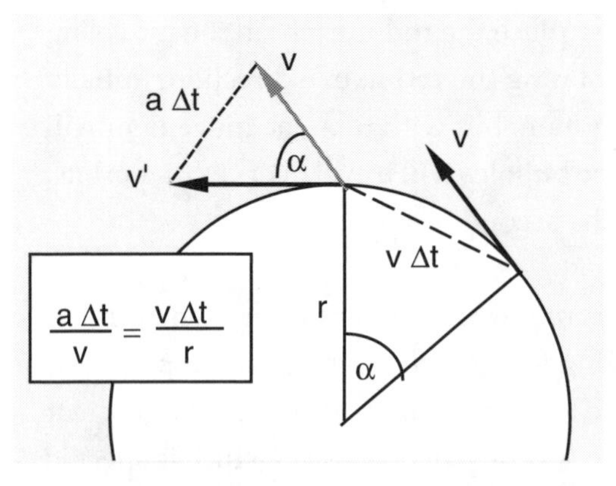

*The centripetal acceleration **a** due to changing tangent velocity v (Δt is a small time interval)*

Christiaan Huygens (1629-1695)

Christiaan Huygens was born on 14 April 1629, in The Hague, Netherlands, to a prominent Dutch family. He was educated by his father and private teachers. He spoke Latin, Greek, French, and some Italian, studied logic, mathematics, mechanics and geography. He played the viola da gamba, flute and harpsichord. From 1645 to 1647, Huygens studied law and mathematics at the University of Leiden, and was influenced by Descartes. Between 1650 and 1666, he lived in The Hague. During these years, Huygens traveled three times to Paris and London, where he met Pascal and Boyle. In 1656, he invented the pendulum clock. He made valuable contributions in wave theory and optics. He worked with Papin on a *moteur a explosion* and was in regular contact with Leibniz. When the Academie Royale des Sciences was founded by Louis XIV in 1666, Huygens obtained an official status and financial aid. As the Academy's most prominent member, he lived in an apartment in the Bibliotheque Royale. Huygens died on 8 July 1695 in The Hague.

7.14. CENTRIFUGAL FORCE (1)

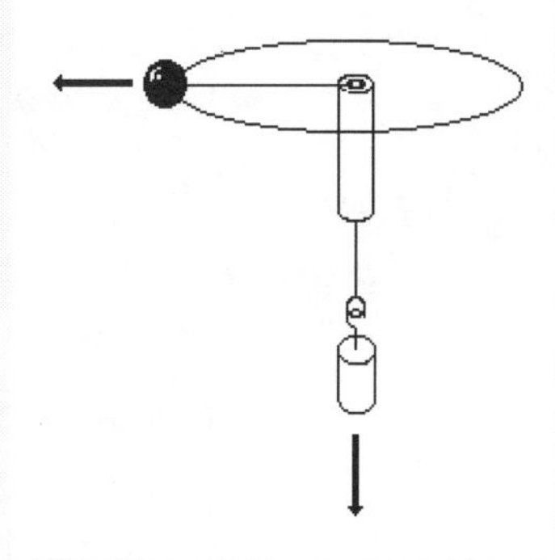

Materials: small bead, weight, tubed handle.

Procedure: Place a small bead and a weight on opposite ends of a string. Thread the string through a tubed handle. When the bead is whirled, it is able to lift the heavier weight.

7.15. CENTRIFUGAL FORCE (2)

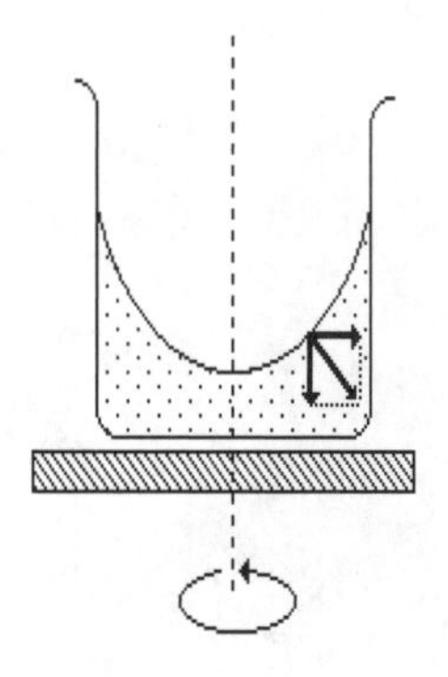

Materials: bucket, water, revolving disk.

Procedure: Place a bucket half-filled with water on a revolving disk. The faster the bucket rotates, the more water is pressed against the wall, causing it to ascend.

Explanation: The surface of the water is always perpendicular to the acting force. In still water, only gravity is present and the free surface is parallel to the Earth's surface. In rotating fluid, the centrifugal force must be added to the gravity force so that the resultant of the two forces determines the orientation of the new surface of water.

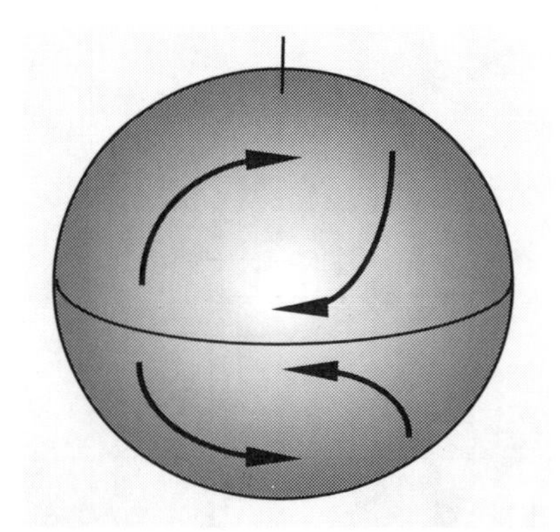

Coriolis effects

CORIOLIS FORCE. In 1835, Gaspard Gustave de Coriolis (1792-1843) published a theory describing the behavior of bodies in motion on a spinning surface. By means of mathematical calculations, he proved that the path of any object moving on a rotating body curves in relation to the rotating surface, as if it were affected by a fictional force (the Coriolis force). On the Earth, the Coriolis force is perpendicular to the object's relative velocity, and is oriented to the right in the Northern Hemisphere, and to the left in the Southern one. Its magnitude is proportional to the product of an object's mass, its velocity, Earth's angular velocity (7.29×10^{-5} radians/s), and sin ϕ, where ϕ is the latitude at which wind occurs.

Gaspard Gustave de Coriolis (1792-1843)

Gaspard Gustave de Coriolis was born on 21 May 1792 in Paris, France. His father was a loyalist of Louis XVI. In 1808, Coriolis entered the Napoleonic Ecole Polytechnique. Later, he served in the Corps of Engineers of the Ponts et Chaussées. In 1816, he became a tutor of analysis at the Ecole Polytechnique. His poor health prevented him from considering marriage. On 6 June 1831, Coriolis submitted a paper to the Academy of Sciences on the existence of a term complementary to relative acceleration. Beginning in 1832, Coriolis assisted Claude-Louis Navier (who derived equations of fluid motion) in teaching applied mechanics at the Ecole des Ponts et Chaussées. In 1836, Coriolis succeeded Navier as a member in the mechanics section of the Academy of Sciences. His major contribution was describing a force in a rotating frame of reference. Coriolis proposed using the one-half coefficient in the definition of *force vive* (kinetic energy), and the term "work" for a force-displacement product.

The Coriolis effect combines two factors, one that exerts its strongest force on objects traveling on a north-south axis, and another which affects objects moving on an east-west axis. The first factor results from the rotational velocity of the Earth's surface, which varies with latitude. A point residing on the Equator moves at a speed of $2\pi R/24h = 463.2$ m/s, while the poles spin but do not move. Hence, air moving north from the Equator begins with larger rotational speed and outruns slower moving portions of the globe. As a result, it relatively curves eastward and ahead of the Earth's rotation. Similarly, air traveling southward, toward the Equator, begins with a low initial velocity and curves west, as the faster-moving Earth exceeds it.

The east-west component of the Coriolis force is a consequence of the tendency of any orbiting object to fly off in a straight line. This tendency together with the rotation of the Earth produces a force which lies on the plane perpendicular to the Earth's axis, and thus has a sideways component in relation to the Earth's surface. Consequently, an object moving east will curve toward the Equator, while a westward object will curve toward the pole.

Deflected trajectories. In 1836, Simeon Denis Poisson (1781-1840) noted that the Coriolis force should deflect a shell fired from a gun to the right in the Northern Hemisphere. At that time, however, this effect was believed to be insignificant. Eighty years later, during World War I, German gunners in March of 1918, observed a peculiar effect during the shelling of Paris. The trajectories of shells fired with a specially built 210-mm howitzer, from about 110 km away from Paris, surprisingly curved to the right and landed about 1 km from the target. A similar observation was made during a British-German naval battle near the Falkland Islands (latitude 50°S), also during World War I. British gunners used the gannery tables valid only for the Northern Hemisphere and were astonished to see that their shots, though well aimed, were mysteriously deviated about 100 m to the left of the German vessels. In both cases, the deflections of projectiles were caused by the effect of the Coriolis force.

7.16. CORIOLIS FORCE (1)

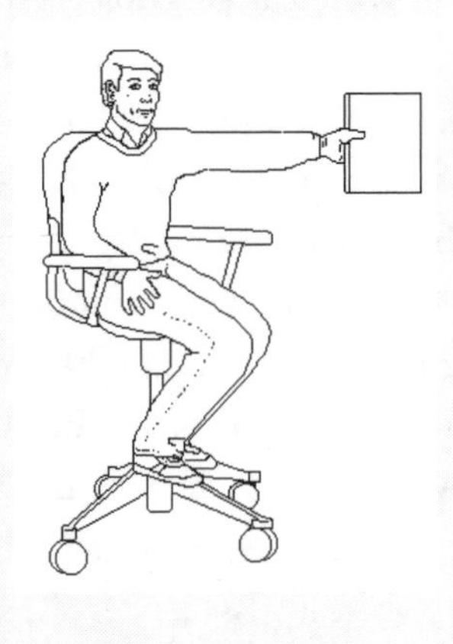

Materials: swivel chair, large, heavy book.

Procedure: Sit in a swivel chair, which is being rotated by another person, and try to extend your hand forward and backward while holding a heavy book. Observe the Coriolis effect due to the conservation of angular momentum.

7.17. CORIOLIS FORCE (2)

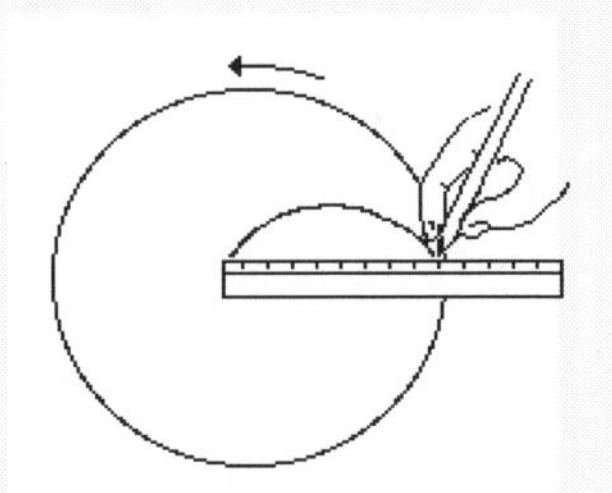

Materials: white sheet of paper, ruler, pencil, pin.

Procedure: Pin the center of a sheet of paper to a table. Using a ruler, draw a line from the center to the edge of the sheet. Try to draw the same line when the sheet is rotated counterclockwise (the help of another person might be needed). Notice that the line is no longer straight.

7.18. CORIOLIS FORCE (3)

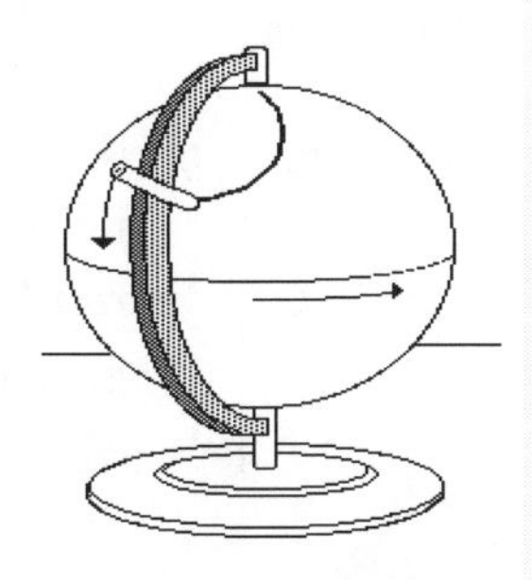

Materials: globe, marker, ruler.

Procedure: Using a ruler, draw a line on the globe from the Pole to the Equator. Try to draw a similar line when the globe rotates counterclockwise. Notice that the line is no longer straight.

The Coriolis force is very small and produces significant effects only for very long-lived phenomena in large weather systems, and in large eddies in the oceans. In short-lived phenomena, the Coriolis force is negligible. This fact might be illustrated using the example of a bathtub vortex. In this case, the Coriolis force does not play any significant role. Disturbances due to asymmetrical initial conditions, such as the influence of temperature, side walls, or already existing fluid motions, determine the clockwise or counterclockwise direction of vortex rotation when water is drained. Only when all disturbances were carefully avoided in laboratory conditions, the Coriolis force would be able to determine the direction of the bathtub vortex.

The Coriolis force often becomes negligible in relation to other acting forces, even in larger-scale processes. For example, even the Coriolis force does probably cause waters of large rivers to press harder on some of their shores, nevertheless, the effects of this action are still vastly exeeded by other acting forces and local circulations. Consequently, the Coriolis force does not cause erosion of river banks and is not responsible for river meandering, as it has been suggested by some authors. The real riverbeds are developed in quite different ways.

EQUATIONS OF MOTION. In 1755, Leonhard Euler (1707-1783) used Newton's second law, and the concepts of partial differential equations, to develop the equations of fluid (gas or liquid) flow for non-viscous fluids. Euler assumed that the resulting acceleration is due to the pressure gradient and gravity forces. Euler's work was the first attempt to formulate equations governing the mechanics of fluid systems.

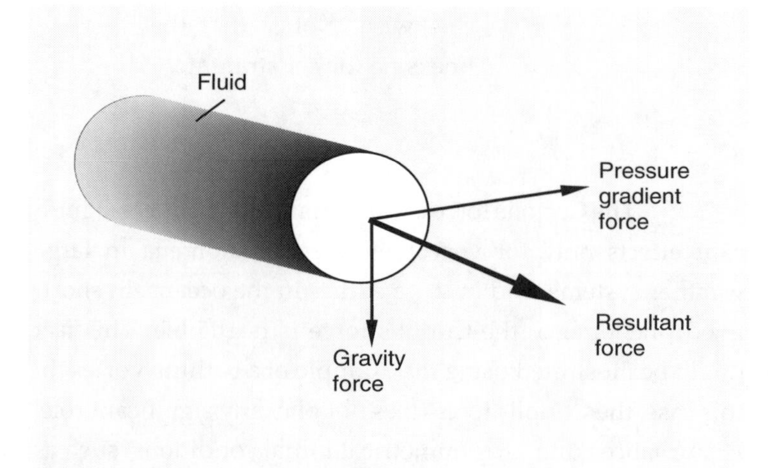

The balance of forces considered by Euler

A debate. Euler was a religious man. Once in St. Petersburg (Russia), he was challenged by the famous French philosopher, Denis Diderot, to a debate on the existence of God. Euler presented his argument in favor of God in the form of an irrelevant algebraic equation. Diderot, who knew no mathematics, was speechless and feeling like a fool quickly left Russia.

In 1827, the significance of Euler's concept was recognized by the French engineer Claude-Louis Navier (1785-1836), and also by his compatriots, Simeon Denis Poisson (in 1831), Augustin Cauchy (in 1841), and Saint-Venant (in 1843), who added the viscous force terms in the above equations of motion. The Englishman, George Stokes (1819-1903), the Lucasian professor at Cambridge (a position previously held by Isaac Newton), addressed a similar problem in 1845. Later, he insisted that his derivation was completed independently, because he was not thoroughly familiar with the French literature on the subject, and also that his assumptions differed sufficiently from those applied previously by French mathematicians. The resulting set of governing equations, commonly used in meteorology, is called the *Navier-Stokes equations.*

Joseph Louis Lagrange (1736-1813) presented a different, but equivalent form of equations (called *Lagrangian* form) in 1760. In 1774, Pierre Simone Laplace (1749-1827) formulated the general equation of fluid motion in both Eulerian and Lagrangian forms, as well as in Cartesian and polar coordinates.

Later, other physical laws were added to the Navier-Stokes equations. As a result, the set of equations governing the fluid motion consisted of six equations with six unknowns. The set included three equations of conservation of momentum, the equation of conservation of energy, the equation of conservation of mass, and the equation of state. It described six unknowns: three components of wind velocity, temperature, pressure, and density.

7.19. CONSERVATION OF MASS

Materials: faucet with running water.

Procedure: Turn a faucet on so that the water runs smoothly. Notice how the water narrows near the bottom of the stream.

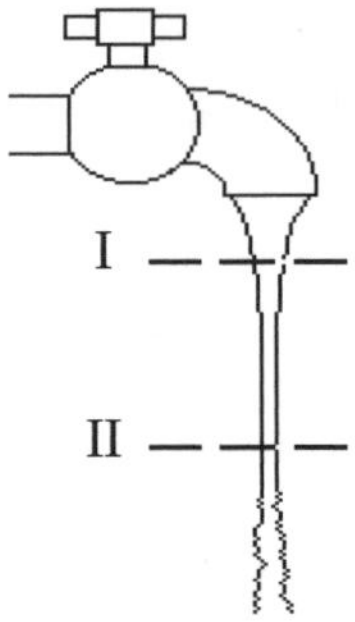

Explanation: According to Galileo's law of falling bodies, the velocity of falling water increases. Consequently, the velocity near the faucet is smaller than near the bottom of the stream. Consider levels I and II, as in the figure. The amount of mass which crosses level I in a unit of time is $\rho A_1 v_1$, where ρ is the water density, A_1 is the water cross- section at level I, v_1 is water velocity at this level. As the mass of water is conserved between levels I and II (no sources, no sinks), the product $\rho A_1 v_1$ must be equal to the amount of mass which crosses level II, which is $\rho A_2 v_2$. As a result: $\rho A_1 v_1 = \rho A_2 v_2$. This equation can be rewritten in the form: $v_2/v_1 = A_1/A_2$, which was derived by Daniel Bernoulli in 1724 (in his *Hydrodynamica*, Daniel Bernoulli stated that *all particles of the liquid in a plane perpendicular to the flow have the same velocity which is inversely proportional to the cross section*). Because $v_2 / v_1 > 1$ (velocity increases) than $A_1 / A_2 > 1$ and stream narrows.

Leonhard Euler (1707-1783)

Leonhard Euler was born in Basel, Switzerland on 15 April 1707. He studied at the University of Basel, a small school with about 100 students and 19 professors. One of his professors was Johann I Bernoulli, who stimulated Euler's interest in mathematics. In 1723, Euler obtained a Masters Degree and joined the Department of Theology at the University. In 1725, the newly organized St. Petersburg Academy of Sciences was searching for personnel. Johann I Bernoulli's sons, Daniel and Nikolaus II, received appointments there. In 1726, as a result of Daniel's persuasion, the invitation was also extended to Euler. On 5 April 1727, Euler left Switzerland (never to return) and arrived in St. Petersburg on 24 May 1727. In St. Petersburg, Euler succeeded Bernoulli (who returned to Basel) as a professor of mathematics, got married, and had five children. In 1740, the King of Prussia, Frederick the Great, made a decision to reorganize the Berlin Society of Sciences (established by Leibniz). Euler was offered a position, which he accepted. On 25 July 1741, he arrived in Berlin and lived there for 25 fruitful years. In Berlin, Euler prepared 380 papers, of which 275 were published. In 1752, he wrote *Principia Matus Fluidorum* (published later in 1761), in which he derived the continuity equation and the general equations of motion. At the request of Catherine the Great (and because of his cold relationship with Frederick the Great), on 9 June 1766, Euler left Berlin for St. Petersburg. On the way, he spent 10 days in Warsaw, at the invitation of the Polish King, Stanislaw August Poniatowski. In St. Petersburg, Euler arrived with his entire family on 28 July 1766. His eyesight was slowly failing (caused by his many observations of the Sun), and by 1767, he was totally blind. Nevertheless, he was still very active in science. In 1773, his wife died. Three years later, he remarried her half sister, Salome. On 18 September 1783, at 5 o'clock in the afternoon, he suddenly suffered a brain hemorrhage. He collapsed uttering: "*I am dying*". He died at 11 o'clock that night. According to Leibniz: "*Euler was glad to observe the flowering in other people's gardens of plants whose seeds he provided*". Euler made important contributions in almost every area of mathematics. He is considered the father of analytical geometry, complex numbers and variational analysis. He invented the number e = 2.7182..., and symbols "i" for $\sqrt{-1}$, and "f" for function. Euler was also interested in the science of meteorology and summarized his meteorological theories in his *Letters a' Une Princesse d'Allemagne* in 1755.

LAMINAR AND TURBULENT FLOWS. In 1883, Osborne Reynolds (1842-1912), a British physicist observed that when water flows slowly out of a container through a long pipe and the flow is marked at the entrance of the pipe with dye, a dye filament at low speed can be seen. Dye particles diffuse slowly and do not have time to spread out. He called such a flow *laminar*. However, when the flow velocity was increased above a critical value of the velocity, at a certain distance from the entrance of the pipe, a sudden change occurred. Reynolds observed in this case disordered movements of single dye filaments. He called such irregular motions *turbulent*.

Reynolds' experiment

Complexity of turbulence. Turbulent flows are considered to be extremely complex and their theoretical analysis is one of the most challenging problems in modern physics. In 1932, the British physicist, Sir Horace Lamb (1849-1934), remarked that when he eventually was taken to heaven, he would like to be enlightened by God on two matters: on quantum electrodynamics, and on the turbulent motion of fluids. He also mentioned of being rather optimistic only about the former.

Reynolds characterized turbulent properties of motion in terms of a number, called the Reynolds number. The Reynolds number is defined as a ratio of the inertial to viscous forcing in fluid motion. When viscosity force dominates, as in very slow motions, the Reynolds number is small. Small Reynolds numbers indicate laminar flows. Conversely, in very fast motions, or low viscosity fluids, the Reynolds numbers are large. Large Reynolds numbers indicate turbulent flows.

Reynolds observed that the transition from laminar to turbulent motion occurs at certain, specified (critical) values of the Reynolds number. Turbulence is generated by thermal and velocity contrasts in the flow. It is dumped by viscosity forces in the process called *dissipatio*n. Dissipation irreversibly transforms the kinetic energy of motion into heat.

Reynolds' apparatus. Reynolds' original apparatus still stands in the hydraulics laboratory of the Engineering Department of Manchester University. Recently, it has been used to repeat Reynolds' experiment. However, vibration from heavy traffic on the streets of Manchester has made the critical value of the Reynolds number substantially lower than the value of 1400 obtained by Reynolds a century ago.

7.20. VERY VISCOUS FLOW

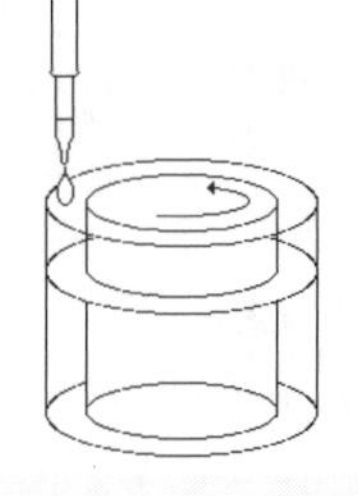

Materials: two circular cylinders of slightly different diameters, syrup, dye.

Procedure: Place syrup between two circular cylinders. Inject a few drops of a dye into syrup. Rotate the inner cylinder while keeping the outer one at rest. If the rotation is stopped after a few revolutions, and the inner cylinder is rotated back, a blob of dye which has been greatly spread in the meantime, will return almost exactly to its original point configuration. This *reversibility* is characteristic of low Reynolds number flows without turbulence.

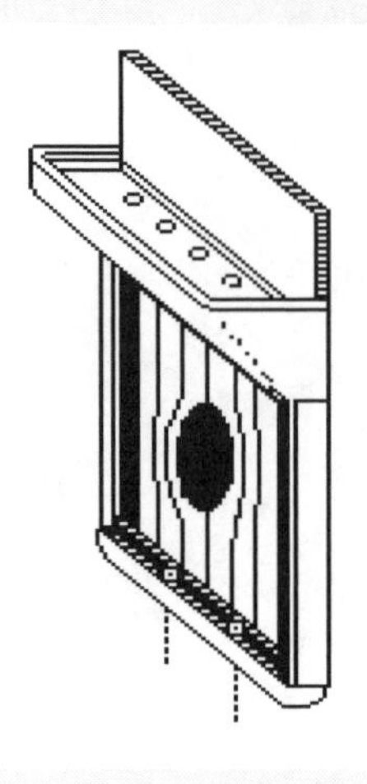

7.21. LAMINAR FLOW

Materials: two rectangular glass plates 20 cm x 30 cm and 20 x 35 cm, waterproof adhesive tape, plasticine, water-soluble glue, crystalline potassium permanganate, sewing needle, cardboard.

Procedure: Cut two straight cardboard strips, 30 cm long and about 1 cm wide, and also a circular disc, 4 cm in diameter. Position the disc in the center of the shorter glass plate. Place the strips lengthwise along the opposite long edges of this plate. Lay the longer glass plate on top, so it extends beyond the shorter one only at one end. Using adhesive tape, bind the plates tightly together along their long sides. Using plasticine, build a large container, where the longer plate extends beyond the shorter one. Seal off the bottom with plasticine. With the needle, pierce two small holes through it. Fasten large potassium permanganate crystals at the bottom of the plasticine container. Supply water to the container. As water slowly drips through the lower holes, each potassium permanganate crystal produces a colored line of liquid. The colored lines mark patterns of motion around the enclosed disc.

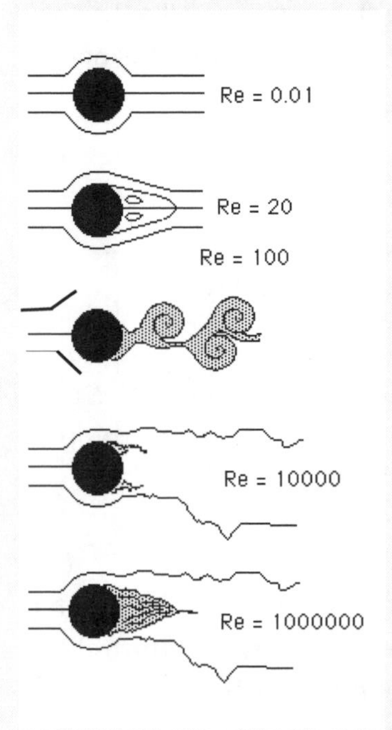

Comment: This flow cell was devised by Hele-Shaw (1898) to display the motion of an effectively inviscid (very small Reynolds number) fluid around obstructions. In this case, the motion around the disc is symmetrical with respect to the disc's axis which is parallel with the flow (see the accompanying figure). For larger Reynolds numbers Re, the symmetry disappears and wakes of various types are formed. Flows with very large Reynolds numbers cannot be generated in the apparatus described above.

7.22. TRANSITION TO TURBULENCE (1)

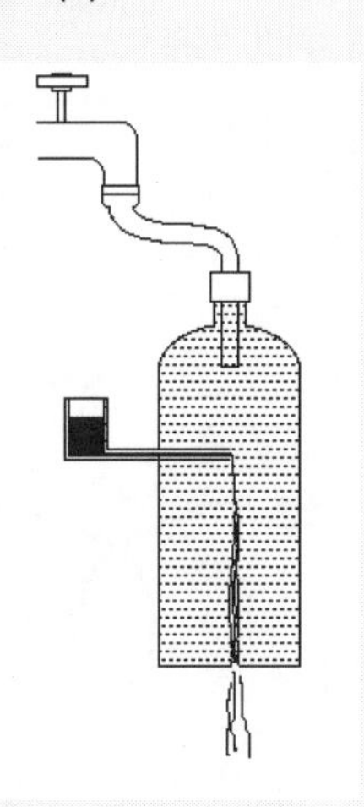

Materials: plastic bottle, long plastic tube, short tube, ink.

Procedure: Connect a plastic bottle filled with water to a faucet by a long tube. Puncture two small, round holes in the bottle, one in its bottom, and one on the side. Water can drain out of the bottle through the hole in the bottom. Insert another tube into the bottle through the hole on the side, and glue it to the face of the bottle. Fill it with ink. Observe ink traces in the bottle.

Comment: If the speed of the flow is slow, the ink trace is narrow, straight and directed along the axis. Such a regular flow is called *laminar*. When the speed of the flow is increased, the ink trace becomes irregular, wavy and finally dyes the interior of the bottle. Such a flow is called *turbulent*.

7.23. TRANSITION TO TURBULENCE (2)

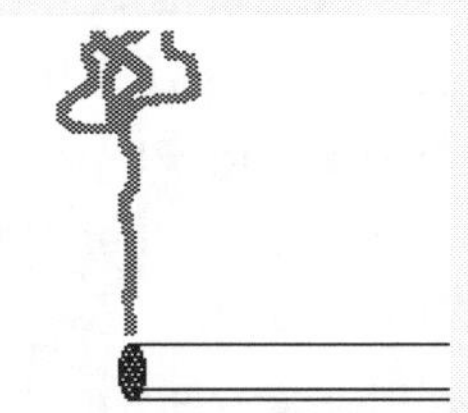

Materials: cigarette.

Procedure: Observe the smoke stream of a lit cigarette. Near the cigarette, the smoke rises smoothly. Farther above it, the smoke forms swirls.

Explanation: Initially the flow of the hot smoke is slow and laminar. When the upward velocity increases, the character of the flow changes from laminar to turbulent.

7.24. DISSIPATION

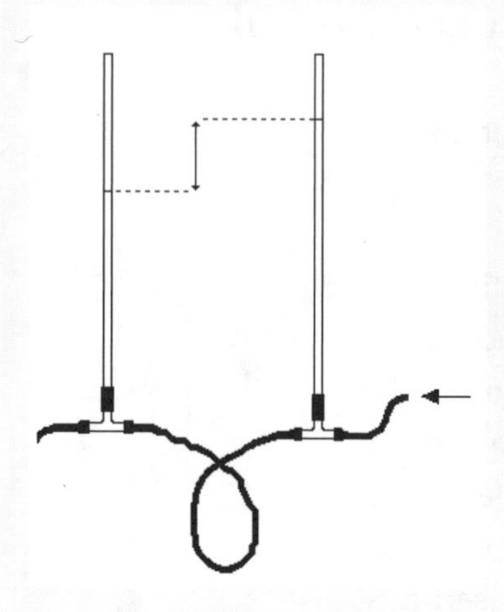

Materials: two rubber hoses of about 0.5 cm bore, one 75 cm and the second 2 m, two glass T-pieces to fit the hoses, two 75 cm clean glass tubes, laboratory stands, adjustable water supply, clamps.

Procedure: Connect the hoses and glass tubes using the T-pieces so that the tubes are vertical and the cross-bars of the T-pieces are at the same level. Use the shorter hose as an outlet and connect the other T-piece to an adjustable water supply. The water levels in the vertical tubes rise as the water is turned up and fall as it is reduced. Notice that the water level is higher in the upstream vertical tube, and decreases downstream.

Explanation: The difference in levels between the vertical tubes indicates that the pressure in the fluid entering the upstream end of the pipe is greater than that at the downstream end. Hence, there is a pressure force acting along the hose between the T-pieces. Because the flow is steady, according to Newton's law, an equal and opposite balancing force arises in the hose. This force is caused by viscous friction between the hose and the moving water. The viscous friction transforms a portion of the kinetic energy of the flow into heat in an irreversible process called *dissipation*.

7.25. DIFFUSION (1)

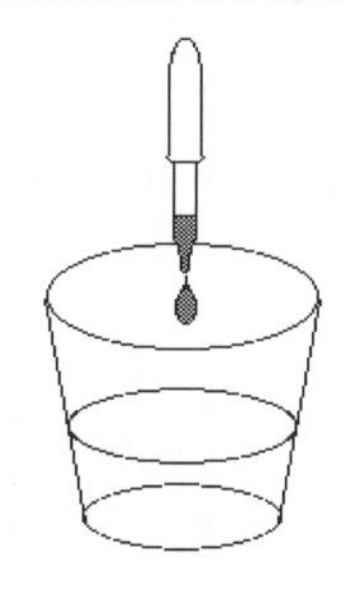

Materials: glass of water, ink.

Procedure: Add a few drops of ink into a glass of water. Do not stir. After a few minutes the ink will be thoroughly mixed in the water. Repeat the experiment by stirring the liquid just before adding the drops of ink. Notice that in the latter case, the mixing process takes place more quickly.

Comment: Diffusion is a process in which heat or mass is distributed in a fluid by its motions. In the first case, the observed mixing was a result of laminar flow and molecular diffusion. In the second case, stirring the liquid in the glass caused turbulent motion and turbulent diffusion. Turbulent diffusion is much more intense than molecular diffusion.

The intensity of turbulence in the atmosphere can vary. This can be illustrated by considering the shape of smoke plumes emitted from a single chimney. The appearance of a plume offers considerable information as to how the thermal and dynamic state of the lower atmosphere influences the turbulent mixing (see the figure on the next page). For instance, when the atmosphere is stably stratified and winds are not strong (as at night in cloudless winter weather), the resulting turbulence is very weak. As a result, a plume expands very little in depth and can be transported by wind over very long distances. On the other hand, for the unstable thermal stratification (which is frequently present by midday) turbulent mixing is very intensive. Large vertical eddies bring the plume to the ground or lift it upwards. Consequently, a plume has a discontinuous shape and is quickly diffused.

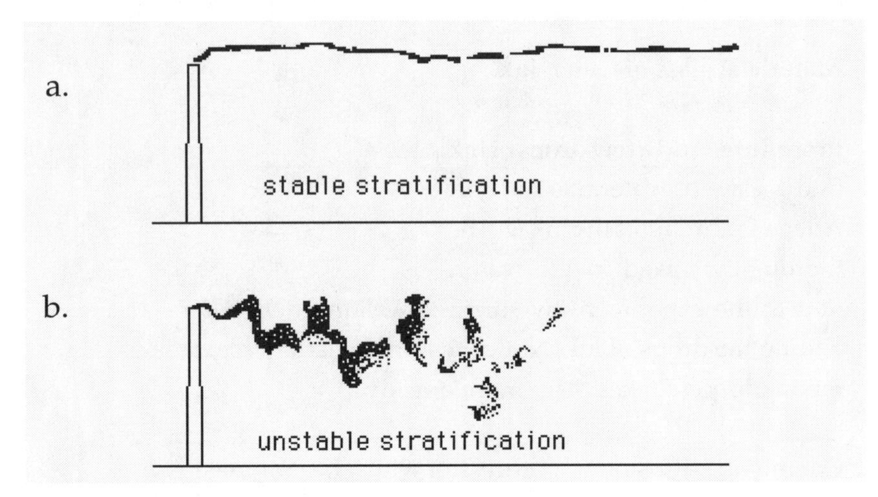

Typical smoke plumes for various turbulence conditions: a. weak turbulence, b. intense turbulence

7.26. DIFFUSION (2)

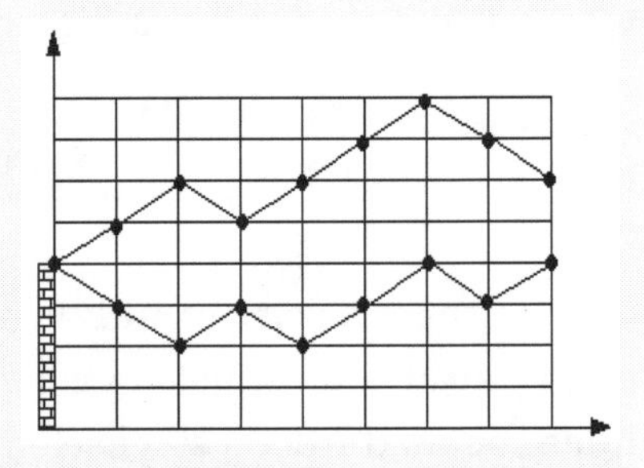

Materials: one coin

Procedure: Simulate diffusion of particles emitted from a chimney by using a grid, as shown in the figure. Imagine that a particle moves in short time steps. On each time step, it is shifted horizontally with the wind by one grid point to the right. At the same time, the particle randomly moves by one grid point up or down. The vertical move is chosen by tossing a coin: upwards - if it is heads, or downward - if it is tails. Perform 8 time-steps for each particle and mark its trajectory. Assume that the particle will be reflected when it reaches the upper or lower boundary, so order is reversed. Repeat the procedure for a few particles. The process can be easily programmed on a computer. A plot of many trajectories looks like a smoke plume.

Everything flows. There is nothing permanent except change. (Heraclitus)

CHAPTER EIGHT

MOTION

Wind Observations, Atmospheric Motion, General Circulation of the Atmosphere, Pressure Systems, Synoptic Meteorology, Polar Front Theory, Pressure Field and Wind, Vorticity, Tornadoes, Convection in the Atmosphere, Atmospheric Boundary Layer, Hurricanes, Can Weather Be Predicted? Numerical Weather Forecasting, Modern forecasting tools.

WIND OBSERVATIONS. The two measurable characteristics of wind are its direction and speed. Observations of wind directions were already made in antiquity. The Babylonians used eight wind directions: four principal and four intermediate ones. The Greeks and Romans used twelve wind directions: four principal and eight intermediate ones.

Aristotle recognized the following winds (depicted in the figure): α. *Boreas*, β. *Meses*, γ. *Kaikias* , δ. *Apeliotes*, ε. *Euros*, ζ. *Phoenicias*, η. *Notos*, θ. *no wind*, ι. *Lips*, κ. *Zephyros*, λ. *Argestes*, *and* μ. *Thraskias*. The white marble Tower of the Winds in Athens still shows the images of eight Greek winds. The 12 m high tower was founded by Julius Caesar (100-44 B.C.) about 50

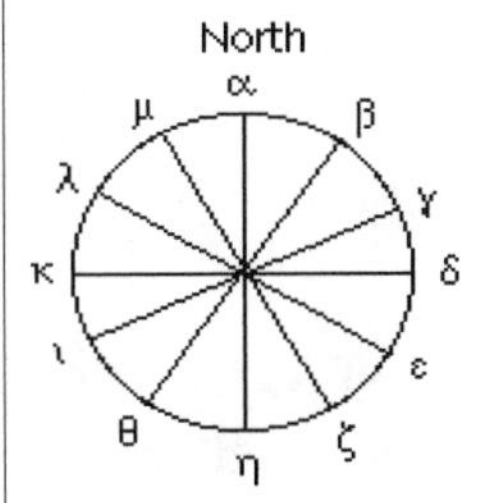

Aristotle's winds

Tower of Winds in Athens

B.C. It was built near the Parthenon by architect Andronicos of Kyrrhos to house a water clock. Above the tower, a bronze statue of god Triton with a rod in his hand showed wind directions.

According to Aristotle's *Meteorologica,* the winds *Thraskias* and *Argestes* were those which brought about clear weather. The winds identified as *Argestes* and *Euros* were initially dry, but turned moist later on. *Meses and Boreas* signalled cold weather, while *Notos, Zephyros* and *Euros* ushered in hot weather. *Kaikias* covered the sky with thick clouds, while clouds brought by *Lips* were scattered. The Romans identified twelve wind directions, which they called: *Serpentrio* (Northerly), *Aquilo, Caesias, Subsolanus* (Easterly), *Volturnus, Phoenix, Auster* (Southerly), *Libonothus, Africus, Favonius* (Westerly), *Corus,* and *Cinicius.*

The measurement of wind speed is a more recent accomplishment. The first *anemometer*, an instrument measuring wind speed, consisted of a swinging plate. It was described by Leon Baptista Alberti (1414-1472) around 1450. Similar instruments were later constructed by Robert Hooke in 1664, and Gottfried Wilhelm Leibniz in 1684.

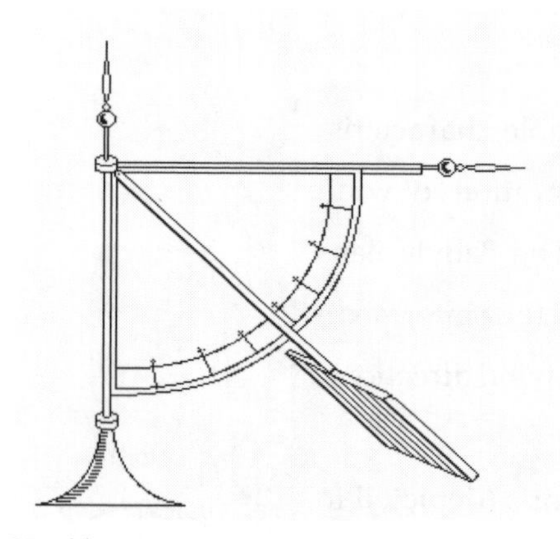

Hook's anemometer

WEATHER FACTS. The highest wind gust on Earth, 373 km/h (231mi/h), was detected on 12 April, 1934 at the 1910-m summit of Mount Washington, New Hampshire. The 5-minute average at the same time was 303 km/h (188 mi/h).

ATMOSPHERIC MOTION. The energy which generates winds and forces atmospheric air to move is supplied by the Sun. However, this fact was not understood until two and a half centuries ago. For instance, the famous French scientist, Jean le Rond d'Alembert (1717-1783), argued that the Sun's radiation affected only the surface, and could not be the primary cause of winds.

In 1746, d'Alembert won a prize offered by the Berlin Academy of Sciences for providing the best mathematical discussion of the atmosphere. The jury was headed by Leonhard Euler (1707-1753). In his work, *Reflection sur la Cause Generale des Vents,* dedicated to Frederick the Great, ruler of Prussia, d'Alembert introduced *partial derivatives* (derivatives for multivariable functions), and also made the first attempt to formulate general equations of atmospheric motion. However, his conclusions were incorrect. He argued, for instance, that the motion of the atmosphere was caused by the tidal effects due to attracting forces of the Sun and the Moon.

Jean le Rond d'Alembert (1717-1783)

Jean le Rond d'Alembert was born on 17 November 1717, in Paris. He was the illegitimate son of Madame de Tencin, a famous salon hostess and the Chevalier Destouches-Canon, a cavalry officer. The infant was abandoned by his mother on the steps of the St. Jean-le-Rond Church in Paris, thus explaining his unusual Christian name. His father, however, located the baby, placed him in the home of the artisan family Rousseau, and provided his son's education. Although he never married, he lived with his mistress, Julie de Lespinasse, for a number of years. d'Alembert became a leading figure in French Enlightenment and the co-editor of the renowned *Encyclopedie.* He published works on calculus, mechanics, music, philosophy and astronomy. Although slight in stature and distinguished by a high-pitched voice, he was known for his wit and gift of conversation. D'Alembert died on 29 October 1783.

In 1926, Lewis F. Richardson noted that atmospheric motion occurs over scales, ranging from thousands of kilometers to millimeters. The largest scale motion, or *general circulation*, is caused by thermal and pressure contrasts over the globe, and modified by the rotation of the Earth. Land and oceans introduce additional modifications to this primary flow and help to initiate secondary circulations. Local topography introduces tertiary circulations. A *cascade process*, in which eddies of the largest (global) size trigger smaller and smaller (local) ones, continues down to molecular motions, which finally cease due to viscosity. Richardson's poetic version depicts the changes quite accurately: "*Big whirls have little whirls, that feed on their velocity; and little whirls have lesser whirls, and so on to viscosity*".

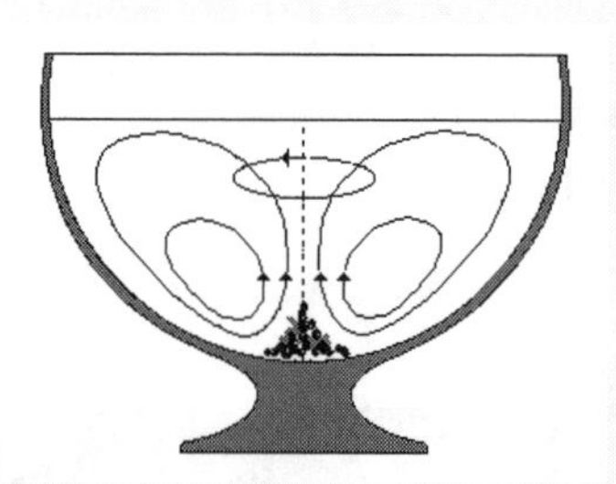

8.1. MOTIONS IN A CUP OF TEA

Materials: cup, tea leaves, water, spoon.

Procedure: Stir tea leaves in a cup and observe their motion.

Explanation: The main body of the fluid is set in rapid rotation, which creates the centrifugal force. Consequently, fluid piles up near the rim of the cup, which creates the pressure gradient. The centrifugal force is almost balanced by a radially inward pressure gradient force. The balance is disturbed near the bottom of the cup due to friction, which causes the fluid to rotate less rapidly. As a result, the pressure gradient force near the bottom pushes the fluid inward (as evidenced by the way the tea leaves in the middle congregate on the bottom of the cup), and eventually upwards.

Comment: A similar phenomenon was observed by the Greek philosopher Empedocles (500-430 B.C.). Empedocles noticed that when a wine rotates in a cylinder, a secondary circulation towards the center is generated. Wine sediments within the fluid migrate toward the center and accumulate there, as long as they are heavy enough to withstand updrafts. Empedocles' explanation of the phenomenon was amusing. He maintained that near the bottom, love brought the liquid together, while at the surface, strife separated it.

GENERAL CIRCULATION OF THE ATMOSPHERE. The largest scale atmospheric motions were first explained by Edmund Halley (1656-1742). In 1686, after a two-year stay on the Island of St. Helena, he formulated a theory about tropical winds. According to this theory, near the Equator, the Sun's heat raises the air and the winds flow to replace it. Halley thought that tropical winds flowed from East to West, following the movement of the Sun.

Edmund Halley (1656-1742)

Edmund Halley was Britain's Royal Astronomer from 1720 to 1742. In 1705, he predicted the return of the comet which bears his name (the comet did indeed return in 1758, 16 years after Halley's death). He was a friend of Isaac Newton, and the financial backer for the publication of Newton's *Principia.* Halley made numerous contributions to the science of weather, the greatest of which was his memoir on the causes of tropical trade winds and monsoons, illustrated with the first meteorological map. Halley based his conclusions partly on the observations he made during his trip to catalog the stars of the Southern Hemisphere, and partly on pieces of information he collected from globe-trotting sailors.

Halley's model was too simplistic, and fifty years later was modified by George Hadley (1685-1768), a lawyer by profession, who was also dedicated to the natural sciences. In 1736, Hadley presented a paper on tropical winds to the Royal Society of London, in which he argued that hot air rose at the Equator, moved towards the Poles, where it cooled and sank, and finally was drawn back towards the Equator. This cell-like cir-

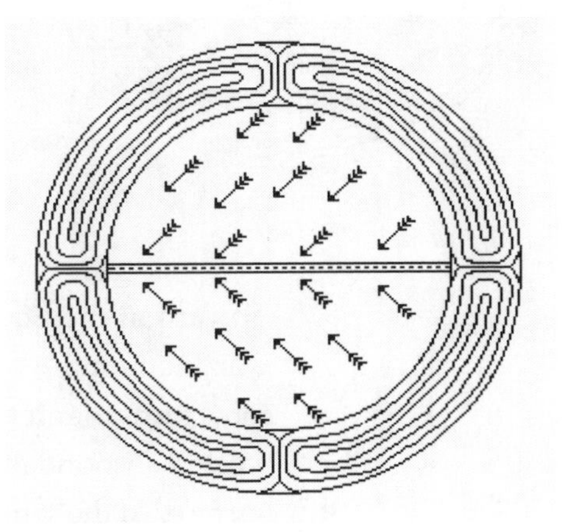

Hadley's model of general circulation

culation was to be present in both hemispheres. Because the Earth rotated eastward, the atmosphere moved along with it. Since the velocity of a point on the Earth was the greatest at the Equator, the air which flowed southward from the Poles lagged behind the Earth's surface and appeared to arrive from an easterly direction.

In 1853, James Coffin (1806-1873) published a monograph *Winds of the Northern Hemisphere*, which presented a series of wind charts from 579 stations of the Northern Hemisphere. Coffin demonstrated an existence of three latitudinal pressure belts and three vertical circulation cells. Several years later, the Russian climatologist, Alexander Voeikov (1842-1916), presented worldwide data from 3223 stations, including ship records compiled by the United States Navy's chief hydrographer, Lieutenant Matthew Fountaine Maury (1806-1873), best known for his charting of the Gulf Stream. These empirical studies provided material for the theoretical work of an American, William Ferrel (1817-1891).

In 1855, based on an analysis of a large amount of information on global winds, ocean currents, and pressure systems, Ferrel found that stable pressure zones existed near the Equator, as well as at latitudes of 30° and 60° on both hemispheres. As a result, he introduced a modification of Hadley's theory by proposing a three-cell hemispherical circulation.

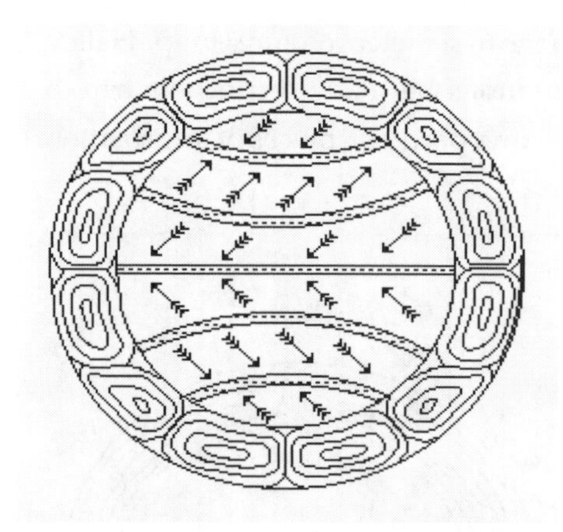

Ferrel's model

The modern picture of the general circulation of the atmosphere is described in the figure on the next page. Air is forced upward at the Equator and begins its high-level flow northward and southward. In the Northern Hemisphere, the Coriolis force turns the air to the right. As a result, upper winds become westerly around the 30° latitudes, and easterly near the Earth's surface. At the same time, the air over the North Pole begins its

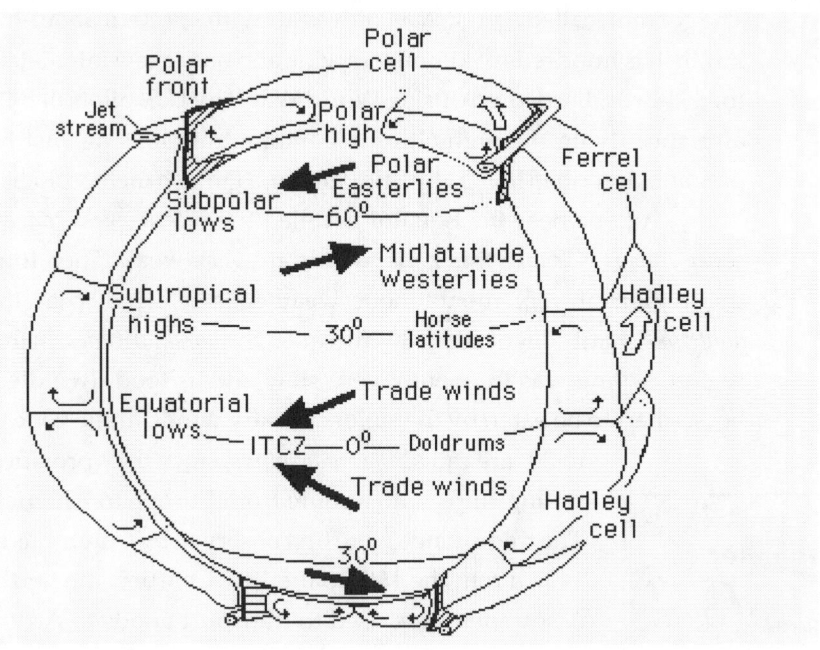

General circulation of the atmosphere (dark arrows indicate winds near the Earth's surface)

low-level journey southward. Likewise, this air is deflected to the right, and becomes easterly around 60° latitude. A similar picture occurs in the Southern Hemisphere.

A semi-permanent high pressure belt (*subtropical highs*) is formed near the 30° latitude in both the Northern and Southern Hemispheres. Air is pushed up around 60° latitude in both hemispheres. As a result, a *subpolar low* pressure zones are created in both hemispheres. A simple convective transfer between the Equator and the Poles is disrupted, and three separate circulation cells are established. The subtropical cell is called the *Hadley cell*, the middle one is called the *Ferrel cell*, and the third one is called the *Polar cell*.

Around 60° latitudes on both hemispheres, a distinct boundary separates cold air moving south, and mild air traveling poleward. This boundary between the Polar and Ferrel cells is called the *polar front*. In the tropopause, above the polar front, a meandering globe-circling current of westerly winds is located.

The current, called a *jet-stream*, moves air with speed of about 150 km/h. It is hundreds of kilometers wide and only a few kilometers thick. It was discovered during World War II by U.S. pilots of B-29 airplanes flying the high-altitude bombing missions against Japan, and described in 1951 by Eric Palmén, a Finnish meteorologist.

A zone near the Equator is called the *Intertropical convergence zone* (ITCZ). In this zone, winds are very weak. Therefore, this region of very monotonous weather is referred to as the *doldrums*. Latitudes of about 30° are called the *horse latitudes*. In this region, sailing was frequently very slow and as food dwindled, horses had to be eaten by the sailors. Steady winds in the zone 0° to 30° are called *the trade winds*, since they provided sailing ships with a route from Europe to America. The trade winds were first observed by Columbus in 1492. From the 16th to the 19th Century, the northeast trades were used to transport goods to Africa, where they were exchanged for slaves. From Africa, sailing boats filled with human cargo voyaged to America, employing southeast trades. From America, with the help of prevailing westerlies, they returned to Europe loaded with sugar, rum and cotton.

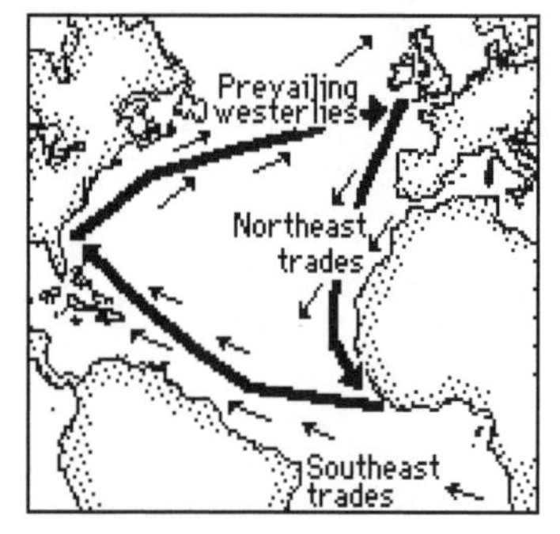

Atlantic trade routes

In the midlatitudes, the flow in the middle and upper troposphere is westerly. Westerlies exist as a result of a balance between the pressure gradient and Coriolis forces. The pressure gradient force is caused by the temperature contrast between the equator and the pole.

El Niño. Normally, the trade winds over the Pacific Ocean blow toward the west. They are accompanied by strong equatorial ocean currents. Every few years, however, the trade winds diminish, while the ocean currents reverse, leading to an eastward flow of warm ocean water along the equator. This causes abnormal weather patterns (heavy rains, floodings, droughts, etc.) observed around the world. The phenomenon is called El Niño (in Spanish "the boy child"), after the infant Jesus, since usually short time reversals of ocean currents along the coast of Ecuador and Peru occur around Christmas.

Westerlies flow in wavelike patterns. Normally, a series of long waves, from 3 to 5 in number, extend completely around the Earth. These waves are called "Rossby" waves, after Carl-Gustav Rossby (1898-1957), who was the first to mathematically investigate this phenomenon. Wave flow of the westerlies provides an important mechanism for heat transfer across midlatitudes.

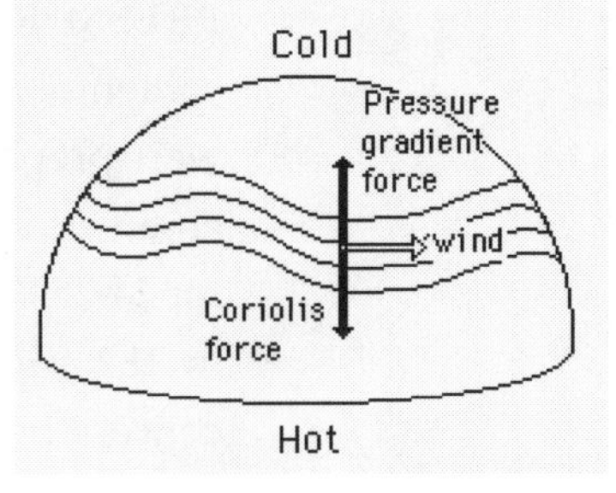

Westerlies

8.2. CIRCULATION MODEL

Materials: a can with a diameter of 10 cm and a depth of 10 cm, 25-cm round and a straight-sided baking pan, ice, water, a few short candles, phonograph turn-table, ink (or aluminium powder).

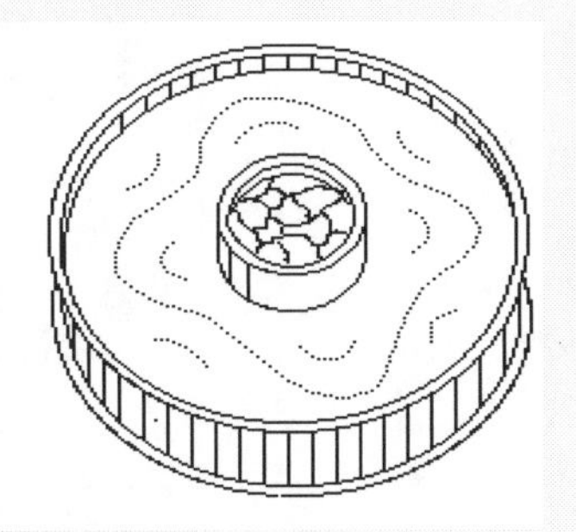

Procedure: Fill the can with crushed ice (to represent the cooling at the Earth's Pole), and place it exactly in the center of the pan. Fill the pan with water to a depth of 4 cm. Begin rotating the turntable. Reduce the speed of the turntable to about 7 revolutions per minute by applying friction. Light candles under the rim of the pan, to represent the heating at the Earth's Equator. Observe the circulation of the liquid in the pan. One or two drops of ink can be added to produce streamers depicting the circulation. The clockwise rotation of the turntable simulates the circulation patterns of the air in the Southern Hemisphere. The circulation patterns in the Northern Hemisphere can be observed in a mirror mounted above the rotating pan.

Comment: After the pan is set into rotation, four stages in cell development can be observed. First, water accelerates slowly to the speed of the pan. Dye streams in the liquid indicate a flow in concentric circles about the ice-filled can. As the effect of heating and cooling becomes apparent, there is a transition from circular flow to long continuous waves. Separate cells can eventually form. The number of waves or cells and their amplitude depends on the size of the pan, speed of rotation and the heat applied at the rim. If the heat source is removed, thermal convection diminishes, and the cells disintegrate.

PRESSURE SYSTEMS. The weather chart is a basic tool used by contemporary meteorologists to analyze weather. The first charts were prepared by the German astronomer, mathematician, and engineer, Heinrich Wilhelm Brandes (1777-1834). Brandes' charts for 24-25 December 1821 and 2-3 February 1823 were published in 1826, in his book *On Rapid Changes in Pressure*. Brandes' concept became generally accepted and widely used by other meteorologists to represent weather patterns on weather charts.

In 1888, a well-known meteorologist, R. Abercromby, in his book entitled *Weather*, presented a classification of pressure fields. His classification is illustrated in the following figure:

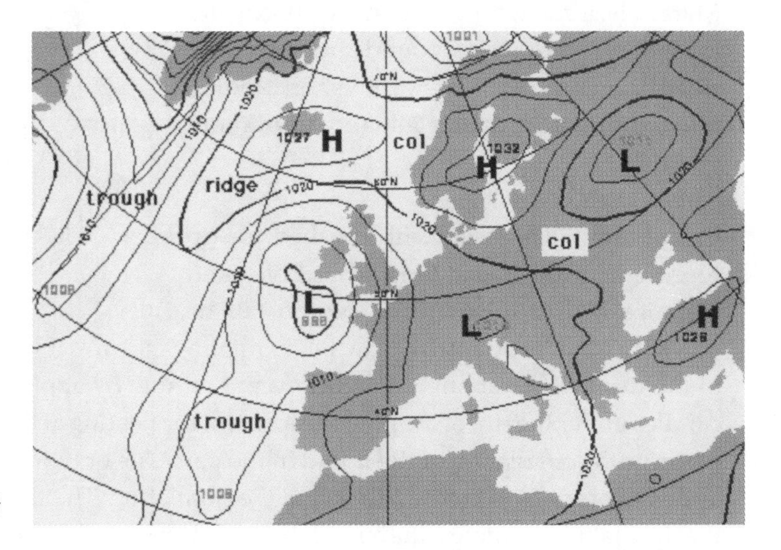

A surface chart

In the figure, contour lines, called *isobars*, connect points of the same pressure. Based on isobaric configurations, five pressure systems are defined as follows:

1. *Low*, also called *cyclone*, a pressure system surrounded on all sides by higher pressure;
2. *High*, also called *anticyclone*, a pressure system surrounded on all sides by lower pressure;
3. *Trough*, an elongated area of low pressure with the lowest pressure along a line marking the maximum curvature;
4. *Ridge*, an elongated area of high pressure with the highest

pressure along a line marking the maximum curvature; and
5. *Col, a* neutral area between two lows and two highs.

The geometrical interpretation of various pressure systems is shown in the figure below. The curved upper surface represents spatial structure of the pressure field at sea level. The lower plane represents its image at a surface pressure chart.

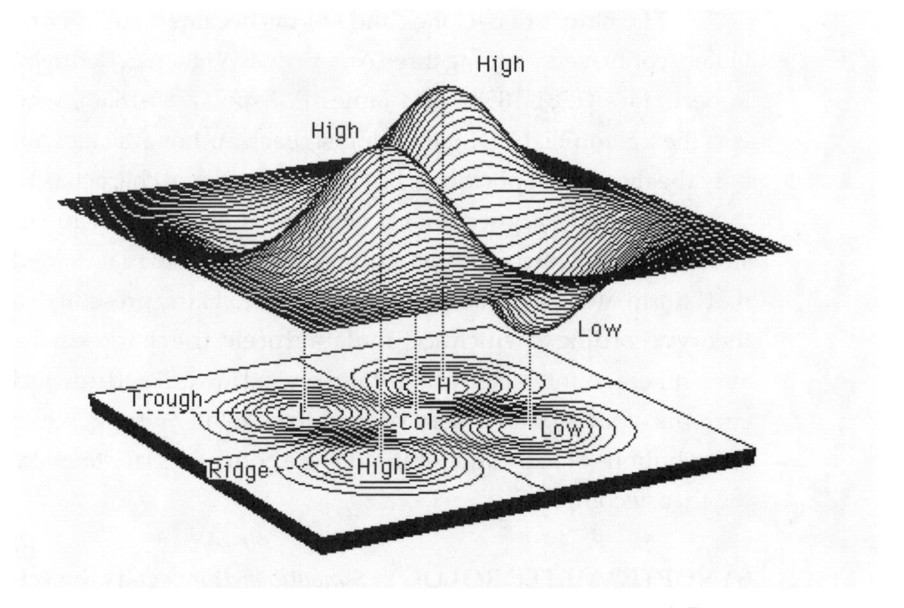

Pressure systems

The word *cyclone*, which Abercromby used in his classification, was invented by Captain Henry Piddington (1797-1858), the curator of the Calcutta Museum, in his *The Sailor's Horn-Book*, published in 1842. The word was derived from the Greek word *kiklos* meaning "circle" or "coil of a serpent", to emphasize the helical character of air motions. The name *anticyclone* was coined in 1863 by the British scientist, Francis Galton (1822-1911).

A characteristic of cyclonic motion was detected in 1821 by an American engineer, William C. Redfield (1798-1857), who noticed that in western Massachusetts the trees blown down after a severe storm (tornado) were pointing just opposite to that which he observed near his home in Connecticut. As a result, he assumed that the storm had been a vortex rotating counterclockwise. He published his observations in 1831.

In 1837, German meteorologist Heinrich W. Dove (1785-1860) formulated a theory of storms, and also developed the idea of a confrontation of polar and tropical currents in the atmosphere. In his *Meteorological Investigations*, Dove related cloudiness, rain and the general character of weather to pressure changes.

The nature of cyclones and storms became a subject of a violent controversy among three Americans: William C. Redfield, Robert Hare (1781-1858), and James P. Espy (1785-1860), who was the first official US meteorologist. Each author attacked not only the theoretical positions, but also the personal characters of the others. Espy argued that a centripetal acceleration forces storm winds toward a low-pressure center. Redfield maintained that storm winds rotate counterclockwise. Hare presented a theory according to which accumulation of electric charges in the atmosphere counteracted gravity and caused inward and upward currents of air. Hare's theory proved to be totally wrong, while contributions of Espy and Redfield were both necessary to make the theory complete.

SYNOPTIC METEOROLOGY. *Synoptic meteorology* is the science of large scale weather processes. The word *synoptic* comes from the Greek word *synopticos* and means "presenting a general view". Theoretical and practical aspects of synoptic meteorology were established at the beginning of the twentieth century in the work of a group of scientists, led by Vilhelm Bjerknes, at the Norwegian Geophysical Institute in Bergen.

In 1904, Vilhelm Bjerknes published a paper entitled, *Weather Forecasting as a Problem in Mechanics and Physics*. In it, Bjerknes planned a procedure that could lead to weather forecasting. He maintained that subsequent atmospheric states developed from preceding ones according to physical laws. To predict the weather, the initial state and the laws of the atmosphere had to be known.

Weather network. Urbain LeVerrier (1811-1877), director of the Paris Astronomical Observatory and a discoverer of the planet Neptune, pioneered the use of the telegraph in a meteorological observation network. After the destruction of the French fleet at Balaclava (the Black Sea) during a storm on 14 November 1854, LeVerrier presented a plan for a meteorological warning system to Emperor Napoleon III, who accepted it. The system consisted of 19 telegraphically linked stations in France and abroad (St. Petersburg, Brussels, Geneva, Madrid, Turin, Rome, Vienna and Lisbon). Similar systems were later organized in Russia (1856), Holland (1859), and Great Britain (1860). Presently, an existing network of meteorological stations consists of 20,000 surface weather stations, and 6700 ships carrying weather measuring equipment. Nevertheless, only 20% of the Earth is considered to be adequately covered.

Bjerknes also realized that a proper description of the state of the atmosphere required global measurements of basic meteorological parameters (pressure, temperature, humidity, and wind). At the beginning of the twentieth century, such a network of meteorological stations did not exist. This situation soon changed. Due to such technological advances as the telegraph (invented by the American, Samuel Morse in 1844), the gathering of weather data became more comprehensive.

Weather services. The first modern weather services were established in the following countries:

1826: Belgium - Observatoire Royal,
1847: Germany - Meteorologische Institut,
1849: Russia - Main Physical Observatory,
1853: Canada - Meteorological Service of Canada,
1854: Great Britain - Meteorological Department,
1855: France - Paris Observatorie,
1870: United States - Weather Bureau.

Because a solution of the system of weather equations was out of the question at the turn of the century, Bjerknes posed the problem of weather prediction in practical form by introducing graphical methods. The foundation of graphical analysis and calculation was laid by Bjerknes and his colleagues in the second volume of *Dynamic Meteorology and Hydrography,* published in 1913.

Vilhelm Frimann Bjerknes (1862-1951)

Vilhelm Bjerknes was born on 14 March 1862 in Christiania (now Oslo), Norway. His father was a lecturer of applied mathematics at the University of Christiania. After completing his studies in Norway, Bjerknes studied classical mechanics and electromagnetism in Paris. He attended the classes of the famous French mathematician Jules Henri Poincare'(1845-1912). From Paris, he moved to Bonn, Germany, and became an assistant to the famous physicist, Heinrich Hertz (1857-1894). In 1902, he switched to geophysics. From 1894 to 1907, Bjerknes was a professor at Stockholm's Hohskola (School of Engineering), and a professor at the University of Stockholm. Between 1912-1917, he worked at the University of Leipzig. In 1917, he returned to Norway. Together with his son Jacob, and his assistants, Halvor Solberg (1895-1974) and Tor Bergeron (1891-1976), he established the Bergen School of Meteorology, which significantly contributed to the development of modern synoptic meteorology. Bjerknes died on 9 April 1951 in Oslo.

In synoptic meteorology, the state of the atmosphere is traditionally represented by a number of charts providing the observed distribution of the meteorological variables from level to level in the atmosphere. *Surface charts,* with isobars on them, depict the atmospheric pressure reduced to sea level. *Upper charts,* with contour lines on them, show heights of constant pressure surfaces of 850, 700, 600, 500, 400, 300, 200, and 100 mb.

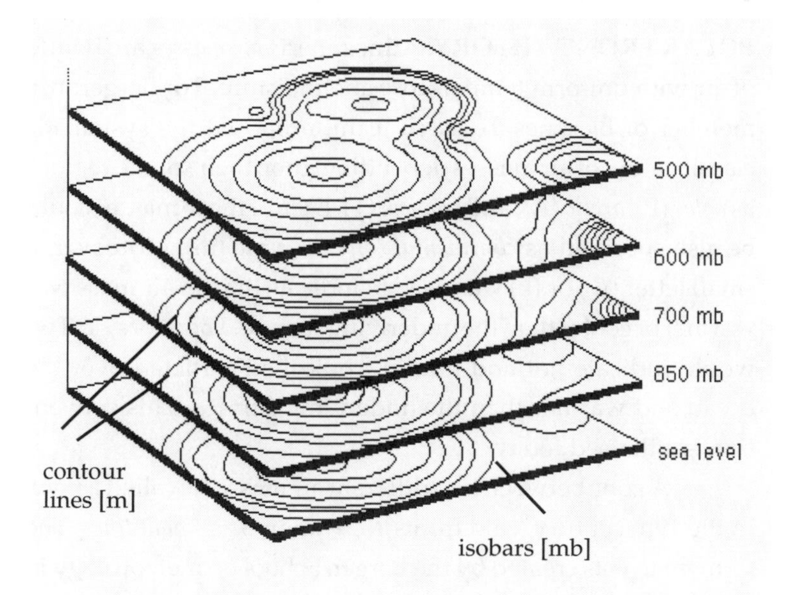

Pressure charts

Absence of the Nobel Prize in Meteorology. One of the most prestigious awards for outstanding achievements in science is the Nobel Prize. Instituted in 1901, the prize is awarded by the Swedish Royal Academy of Sciences, on recommendation of the Nobel Committee. A meteorologist has never won the prize, and meteorology was out of favor with the Nobel Committee. One of the casualties of this policy was Vilhelm Bjerknes, whose candidacy was blocked in 1923, 1924, 1926, 1929, and repeatedly in the 1930s. Behind these rejections was Dr. C. W. Oseen, a member of the Nobel Committee. Oseen claimed that Bjerknes' baroclinic circulation theorems were already implied by H. Helmholtz, and that Bjerknes' theories for polar fronts and cyclones were not confirmed by meteorological phenomena in Sweden. Oseen also maintained that Bjerknes did not understand modern physics. He also thought that meteorology could not be considered a part of physics, hence it was not in the domain for the prize. Nevertheless, Bjerknes' contributions were recognized elsewhere. In 1932, he was awarded the Symons Gold Medal by the British Royal Meteorological Society. In 1940, the same award was given to his son, Jacob.

POLAR FRONT THEORY. Atmospheric *air masses* are bodies of air with uniform temperature and moisture. Tor Bergeron, a member of Bjerknes' research team, proposed the systematic classification of air masses depending upon their source regions as *polar* (P), *arctic* (A), and *tropical* (T). Each of those masses could be also specified as *continental* (c), or *marine* (m). Moreover, a small letter (w) or (k) was used to indicate that an air mass was warmer or cooler than the underlying surface. Therefore, "mTw" would indicate hot and humid tropical air initiated over the ocean and warmer than the underlying surface. This convention is still used today.

A zone between two different air masses is called a *front*. In the 19th century, cold fronts were known as *squall lines*. The term *front* was created by the Bergen School of meteorology in 1920, and was taken from the combat terminology of World War I. Across the frontal zone, temperature, humidity and wind often change rapidly over short distances.

In the 1850s, the American meteorologist, James P. Espy, detected a front line while making up synoptic charts. He called it the "center line of a storm". Another American, William Blasius (1818-1899), also used the concept of a frontal surface, which he called the "surface of interaction". In 1868, Herman Helmholtz had shown that surfaces of discontinuity could exist, and studied the analytical conditions of equilibrium of a surface dividing a warm air mass from a cold one on a rotating Earth.

Fronts can be classified into four groups: *cold, warm, occluded* and *stationary*. The *cold front* is the leading edge of an advancing cold air mass. Analogously, the leading edge of an advancing warm air mass is called the *warm front*. When the cold front catches up with the warm front, the two *occlude* (close together). The result is an *occluded front*. When neither the cold air mass nor the warm air mass replaces the other, the front is called *stationary*.

In 1922, Jacob Bjerknes (1897-1975) and Halvor Solberg (1895-1975) published a comprehensive description of the life cycle of the cyclonic system. According to their model, the

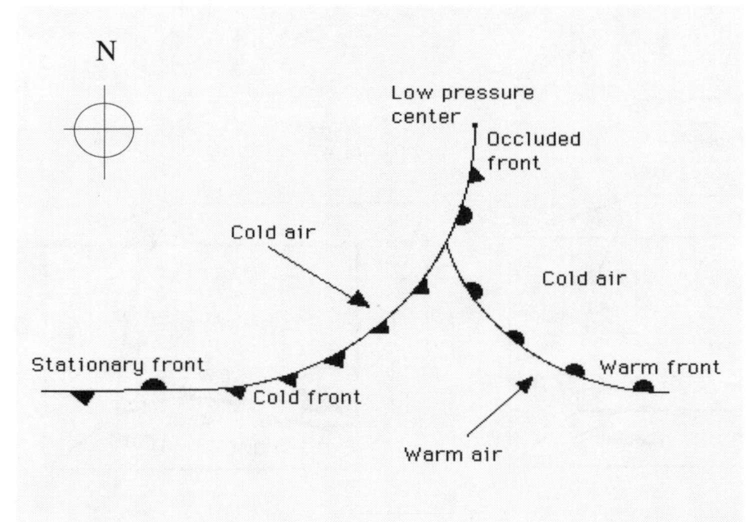

Notation of fronts on weather charts

development of a wave cyclone begins along the *polar front*, which is a stationary front usually occurring around 60° latitude. The stages of a developing wave cyclone are illustrated in the sequence of figures shown on page 248.

The first figure (a) shows a segment of the *polar front* as a stationary front. Imagine that cold air is located to the north and warm air to the south. Air flows parallel to the front, but in opposite directions. Under certain conditions, a wavelike kink forms on the front (Figure b). This leads to the formation of a wave called a *frontal wave*. In the frontal wave, a cold front pushes southward and a warm front moves northward (Figure c). The lowest pressure is at the junction of both fronts. As the cold air displaces the warm air upward along the cold front, a narrow band of precipitation forms. The region of warm air between the cold and warm fronts is called the *warm sector*. The faster moving cold front squeezes the warm sector and eventually overtakes the warm front (Figure d). The system becomes *occluded* (Figure e). At this point, the system is usually most intense but slowly starts dying out. The final dissipative stage is shown in Figure f.

Frontal cycle

a. Polar front stage

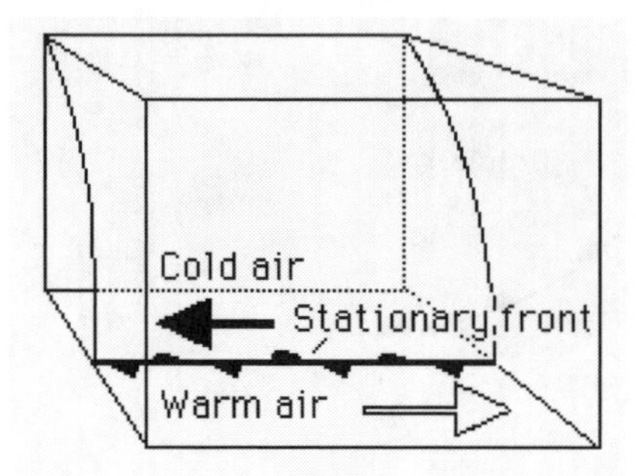

b. Frontal wave formation

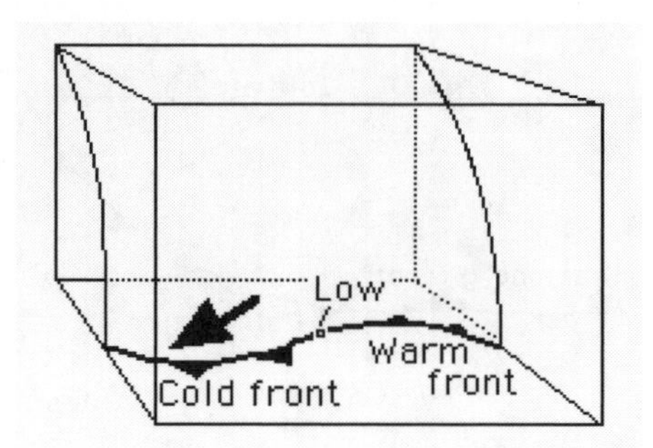

c. Formation of warm and cold fronts

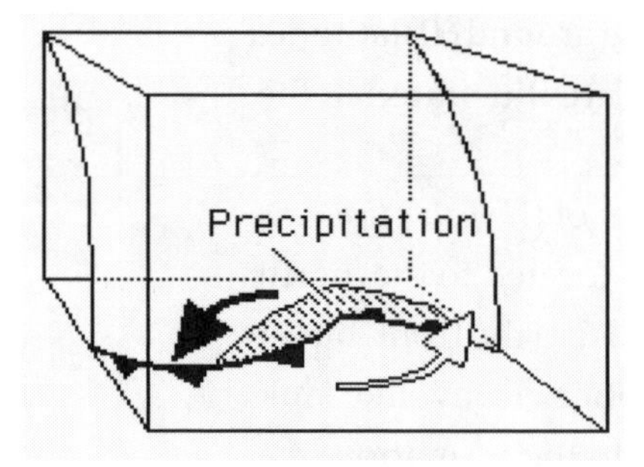

d. Beginning of occlusion

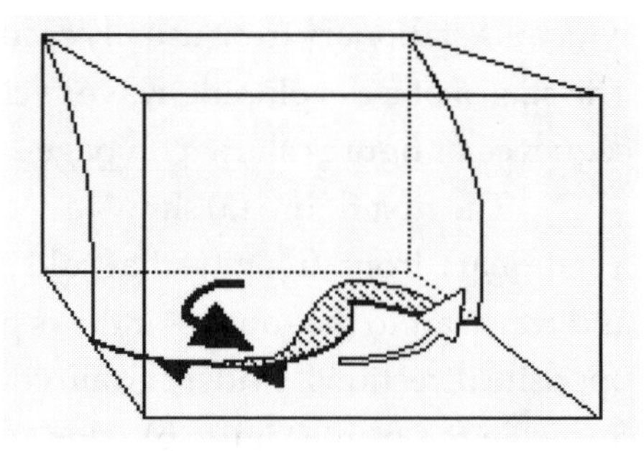

e. Stage of occlusion

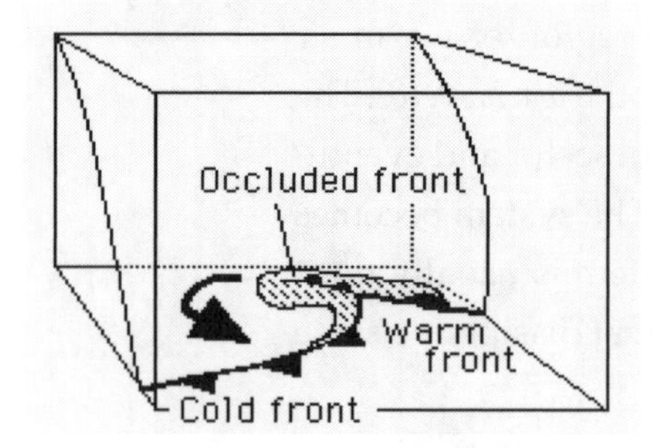

f. Dissipation

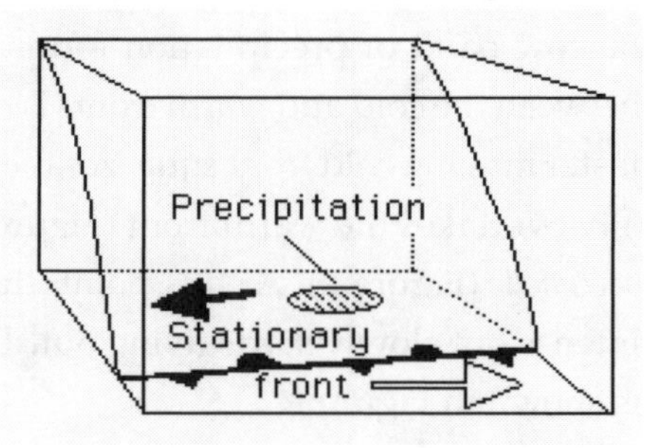

8.3. FRONTAL SURFACES

Materials: 1 m x 0.5 m x 0.25 m aquarium , two 1 mm-aluminum sheets, plastic guides, glue, water, blue and red food coloring, salt.

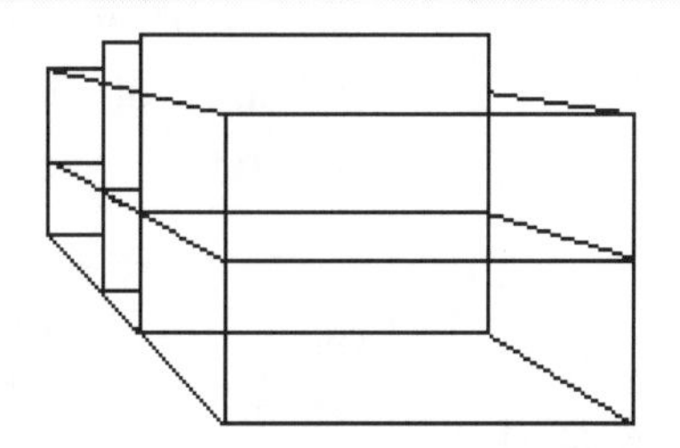

Procedure: Cement plastic guides to the sides and bottom of the aquarium. Place two aluminum partition sheets into plastic guides. Fill the middle section of the aquarium with fresh water to approximately 40 cm deep. Fill the left-hand side of the model, and add 20 tablespoons of salt to it. Stir blue food coloring into this water. Fill the right-hand side of the aquarium with water, and add 10 table spoons of salt to it. Stir red food coloring into this water. In the model, very salty (blue) water represents the cold air mass, less salty (red) water represents the warm air mass, and the fresh water represents an atmosphere separating the two advancing masses. Raise the right-hand partition about 3 cm above the bottom. When the red water reaches the middle of the center section, raise the left-hand partition about 3 cm above the bottom. Observe the interaction of the masses of colored water.

Comment: The interaction of the masses of colored water in the tank is very fast. The interaction of the masses of air in the atmosphere is much slower, due to the fact that air is 750 times less dense than water.

At the cold front, colder and denser air wedges under the warmer air and forces it upward. The frontal edge is 1:50 steep (1 unit of height : 50 units of length). The steepness is due to friction which slows the flow near the ground. As the moist, unstable air rises, its water vapor condenses into a series of cumulus clouds, cumulonimbus (Cb), and altocumulus (Ac). Strong, upper level winds blow the cirrostratus (Cs) and cirrus (Ci) clouds far in advance of the approaching front, as shown in the following figure:

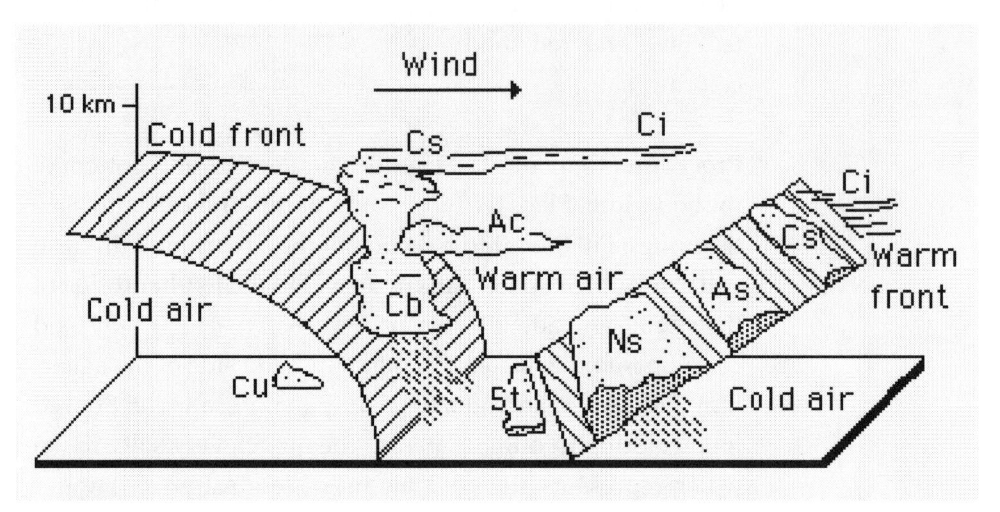

Frontal cloud systems

Warmer air, overriding the colder air, produces a warm front. The frontal edge of the warm front is less steep (about 1:100). Behind a warm front the stratus clouds (St) are observed near the Earth's surface. Stratus clouds can produce drizzle. As air moves upward along the warm front, nimbostratus clouds (Ns) form, yielding a broad area of rain or snow. Farther along the front, clouds gradually transform into altostratus (As), and then into a thin, white veil of cirrostratus (Cs). On top of the frontal surface, there are usually cirrus (Ci) clouds.

Very large raindrops are capable of falling, at the very most, 3 km before they break and evaporate. This indicates that raindrops, produced by clouds with bases higher than 3 km, do not reach the Earth's surface. Consequently, when the frontal surface is sloped as 1:100 (or 3 km : 300 km), the frontal rain band typically extends not more than 300 km ahead of the warm front.

A cold front moves faster then a warm one, and eventually may catch up with a warm front to form an occluded front. Two forms of occluded fronts can be observed in the atmosphere depending on temperatures in occluded masses of air. During the *cold-type occlusion*, the air under the cold front is colder (and heavier) than air under the warm front. On the other hand, during the *warm-type occlusion*, the air under the cold front is warmer (and lighter) and overrides the colder air under the warm front. The cold-type occlusion is more common, but the warm-type occlusion is found frequently in the northern parts of the western coasts of Europe and the Pacific Northwest.

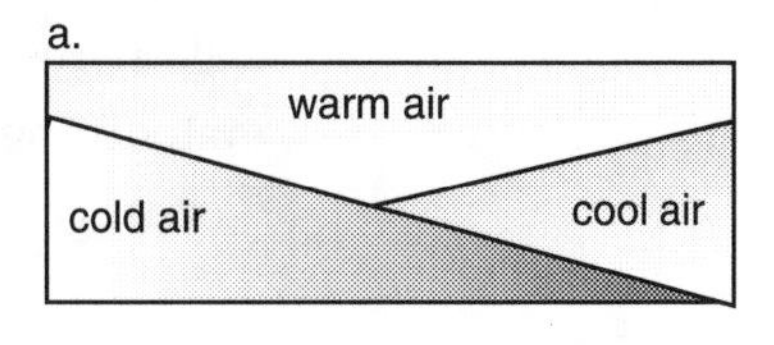

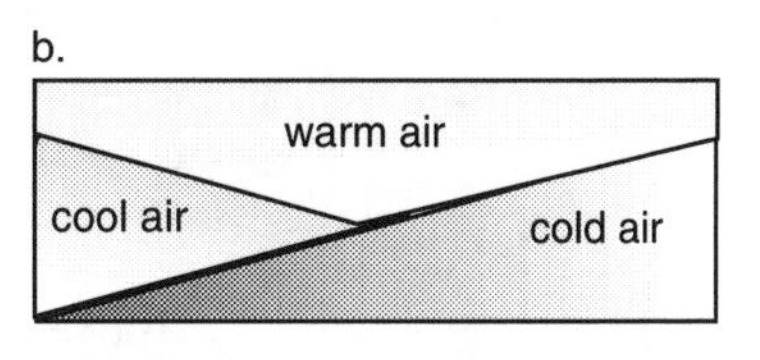

Cold-type (a) and warm-type (b) occlusion

PRESSURE FIELD AND WIND. In 1857, the Dutch meteorologist, geologist, and mathematician, Christoph H. D. Buys-Ballot (1817-1890), introduced a simple empirical rule, known as the *Buys-Ballot law*, which relates the direction of the wind near the Earth's surface to the pressure field. As shown in the figure, if you stand in the Northern Hemisphere, with the wind blowing at your back, the low pressure center will be located to your left. In the Southern Hemisphere, with the wind blowing at your back, the low pressure center will be located to your right. The first theoretical explanation of the rule was given by William Ferrel (1817-1891) about 1859.

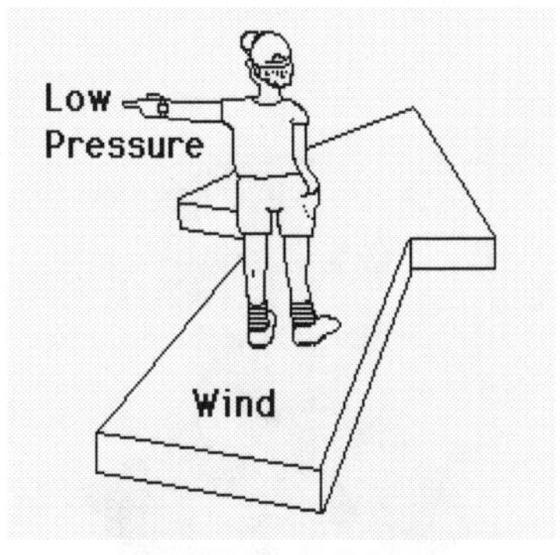

Buys-Ballot's law (the Northern Hemisphere)

Ferrel found that when the air mass is pushed by the horizontal pressure gradient force, it initially moves towards the low pressure area (vector V_1 in the figure on the next page). The moving parcel is simultaneously under the influence of the Coriolis force (outlined arrows indicated as C), which changes its direction (vectors V_2 and V_3). The air changes its direction until an equilibrium between the pressure and the Coriolis forces is reached. The wind resulting from this equilibrium blows along isobars (lines in the figure marked 1000 mb, 1004 mb), and is called *geostrophic*. The term was coined in 1916 by the British meteorologist, Sir Napier Shaw (1854-1945), from the Greek words: *ge*, "the Earth", and *strephein*, "to turn".

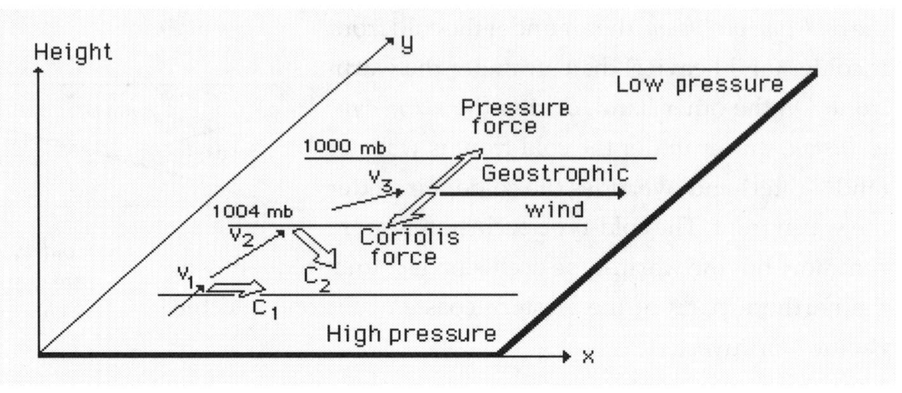

Geostrophic balance

William Ferrel (1817-1891)

William Ferrel was born in Fulton County, Pennsylvania, into a farming family. From 1846 to 1857, as a provincial teacher, he taught school in Montana, Kentucky, and Tennessee. His scientific career began rather late in life, after he became interested in Laplace's work on "Celestial Mechanics". Ferrel made some significant additions to Laplace's work. In 1856, he became acquainted with Lieutenant Maury's scheme of the general circulation. Between 1859 and 1860, he made his own attempt to construct a theory of the general circulation, based on complex hydromechanical equations for the atmosphere of a rotating Earth. In 1859, he independently arrived at the concept of a deviating force due to rotation (Coriolis force). Ferrel studied the heat balance of the Earth' surface, tides, currents, and storms. He also derived the formula for the change of the geostrophic wind with height (thermal wind), and presented the convection theory of cyclones. From 1867 to 1882 Ferrel served as a member of the U.S. Coast and Geodetic Survey. From 1882 until 1886, he was a member of the Signal Service. He died on 18 September 1891 in Maywood, Canada.

Near the Earth's surface, due to friction, the wind velocity is less than the geostrophic wind velocity. In the presence of friction, the pressure force is no longer balanced by the Coriolis force, and the wind is directed from high pressure to low pressure, crossing isobars at an angle of about 10° - 50°. Winds, which blow at altitudes higher than 1 km above the Earth, follow the Buys-Ballot law, and consequently can be considered geostrophic.

In cases with strongly curved isobars, the balance of the pressure, Coriolis, and friction (near the surface) forces yields the centripetal force. The centripetal force in the Northern Hemisphere causes clockwise circulation in anticyclonic (high) pressure systems, and counterclockwise circulation in cyclonic (low pressure) ones. Near the surface, the wind crosses isobars, and is directed from high pressure to low pressure, (see the figure below). The resulting inward motion toward a low pressure center is called *horizontal convergence*. Opposite, outflow in a high pressure center is called *horizontal divergence.*

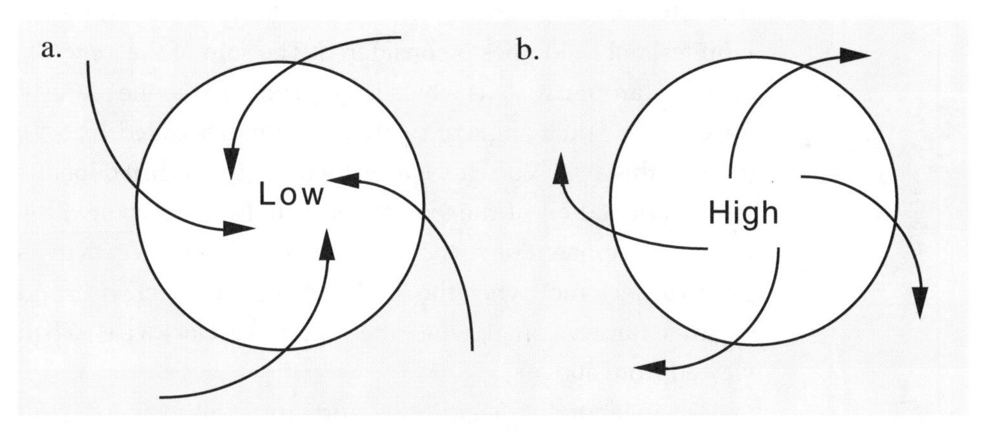

Horizontal convergence (a) and divergence (b) near the surface on the Northern Hemisphere

Horizontal convergence near the ground in the low-pressure system causes the accumulation of air in the center. To remove inward-flowing air, a very slow (a few cm/s) but persistent vertical upward motion is generated, as shown in the figure on the next page. On the other hand, there is a descending flow of air to compensate for the high-pressure divergence near the ground.

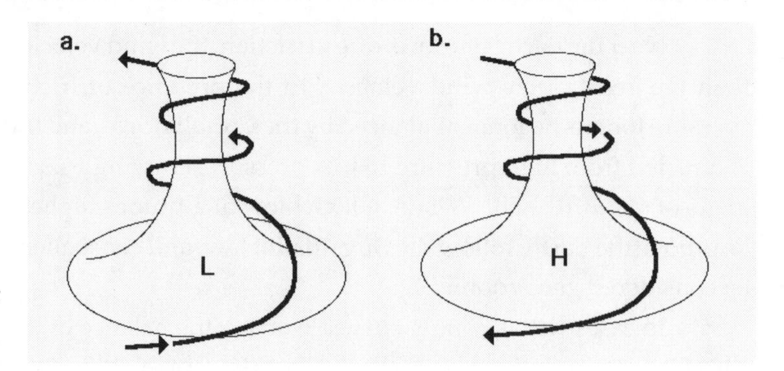

Vertical motions in:
a. a cyclone,
b. an anticyclone

VORTICITY. Leonhard Euler (1707-1753) and Jean le Rond d'Alembert (1717-1783) deserve credit for inventing a mathematical term, which later become known as *vorticity*. Because most fluid phenomena on the Earth involve rotation, the concept of *vorticity* is useful to explain complex atmospheric motions.

Vorticity occurs as a result of different portions of fluid being moved by different amounts. To define vorticity, a cross-like element "+" between two mutually perpendicular infinitesimal fluid lines is considered. The sum of their *angular velocities* (around an axis which is perpendicular to the plane of the cross "+"; unit: angle of rotation per time) is called *vorticity* (around this axis). Vorticity is a vector quantity, since it depends on the orientation of the axis of rotation. In meteorology, the vertical component of vorticity is often considered. Vorticity is positive (cyclonic), when the fluid spins counterclockwise, and negative (anticyclonic), when the fluid spins clockwise (when viewed from above).

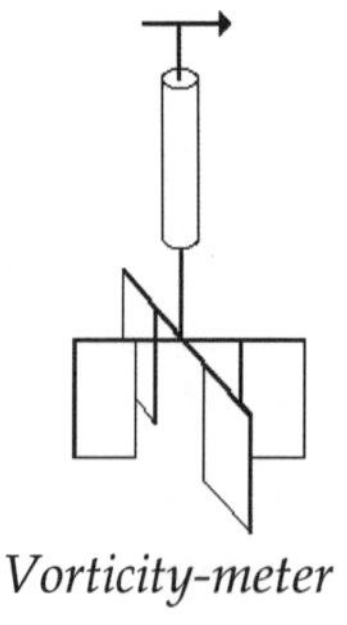

Vorticity-meter

In laboratory conditions, vorticity (around an axis) can be measured by a simple *vorticity-meter*, which consists of four vanes rigidly attached at right angles to a vertical tube. The arrow is affixed to the top of the tube, and rotates with the average angular speed of the pair of perpendicular fluid lines, which coincide with the vanes.

Because the Earth spins, it also has vorticity. In the Northern Hemisphere, the Earth's vorticity is always positive, because the Earth spins counterclockwise about its vertical axis.

The amount of the Earth's vorticity depends on latitude. If the vorticity-meter is placed on the North Pole, it will spin about its vertical axis, with the speed of one revolution per day. Thus, according to our definition, the Earth's vorticity equals the doubled angular velocity of the Earth. When the vorticity-meter is placed on the Equator, it will not spin about its vertical axis. The *absolute vorticity* is defined as a sum of the Earth's vorticity and the vorticity of the air relative to the Earth.

The concept of vorticity is useful for explaining many phenomena in the atmosphere. For instance, it can be used to explain the development of Rossby waves, described on page 239. Consider a parcel of air in the middle troposphere (about 5 km above the Earth) at position "1" in the accompanying figure. Meteorologists discovered that such a parcel conserves its absolute vorticity. Imagine that the parcel is heading toward the pole, to the region of increasing Earth's vorticity. To keep the absolute vorticity constant, there must be a corresponding decrease in the relative vorticity of the parcel. Consequently, at position "2", the parcel turns clockwise toward southeast. Now the air is moving into a region where the Earth's vorticity is smaller. As a result, the parcel's relative vorticity must increase. At position" 3", the air turns counterclockwise, and at position "4" begins to head toward the Pole again.

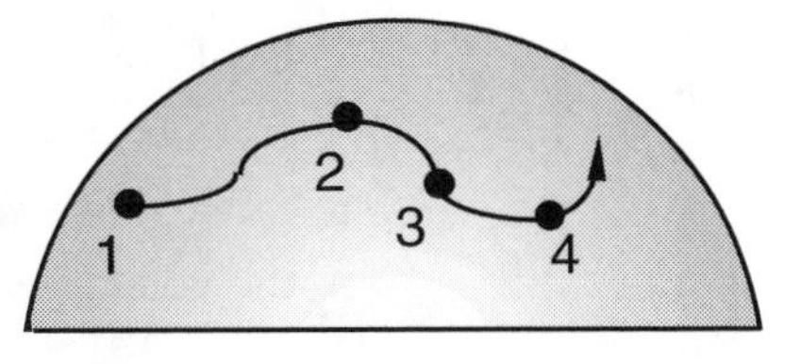

Rossby waves

Next, consider a mass of air which spins counterclockwise, and therefore has a positive relative vorticity. Suppose that the air mass converges and rises in the center. According to the law of the *angular momentum* conservation, a product of the air mass, its velocity, and its distance from a pivotal point is constant. Therefore, as the distance from the center decreases, the velocity of the spinning air must increase. This means that the relative vorticity of the converging air also increases and becomes more positive.

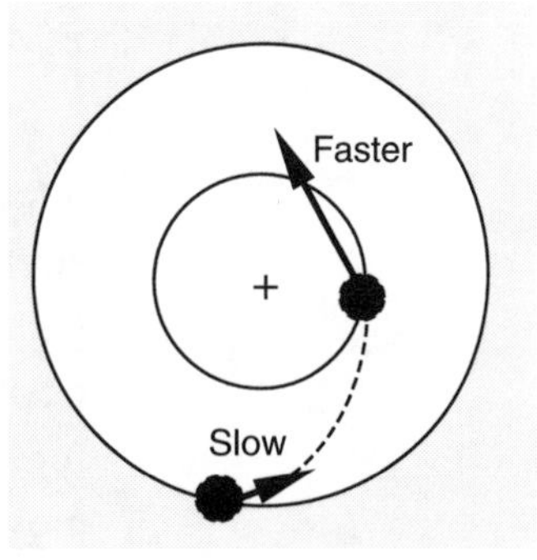

Converging air with a positive vorticity

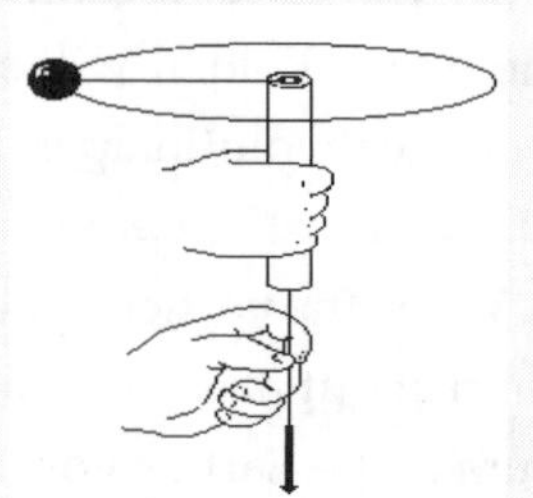

8.4 ANGULAR MOMENTUM (1)

Materials: small bead on a string, short tube as a handle.

Procedure: Attached a small bead to the end of a string threaded through a handle. Twirl the bead. When the length of the string is shortened by pulling it down through the handle, the bead's velocity rapidly increases.

Explanation: The experiment demonstrates the law of conservation of angular momentum, **mvr** = constant. Because mass **m** is constant and the radius **r** decreases, then the velocity **v** must increase.

8.5. ANGULAR MOMENTUM (2)

Materials: spinning chair, bicycle wheel.

Procedure: A person sits in a chair which can spin around its axis. Another person spins the bicycle wheel clockwise and then hands it to the first person. The axis of the wheel is kept horizontal. At this point, the total angular momentum of the chair and the wheel are equal to zero and no motion results. Now, the person sitting in the chair changes the axis of the wheel to vertical position. Because the total angular momentum has to remain unchanged, the chair with the person in it starts moving counterclockwise.

TORNADOES. A tornado is an example of an intense vortex in the atmosphere. It is a local storm accompanied by a violently rotating column of air, usually funnel-shaped, extending down from a cumulonimbus cloud. The funnel cloud serves as a direct marker of the vortex tube and of very low pressure. In a rapidly moving thunderstorm, the funnel cloud can tilt and elongate. As a result, the vorticity intensifies, and hence produces a potentially devastating rotary motion.

Tornadoes occur mainly in the United States, where they are tracked during the entire year, but most frequently during the spring time. The peak activity time usually occurs during the month of April, with an average of 170 cases reported in the United States during this month. Tornadoes are typically produced during severe mid-afternoon or early evening thunder-storms. They travel at an average speed of 50 km/h, but higher speeds have also been reported. Their average path length is 15 km with a width of 200 m. Tornadoes produce loud roars similar to the sounds of airplanes or freight trains. On the average, about 100 deaths per year are caused by tornadoes in the United States.

Tornado

Most damage caused by tornadoes is due to violent winds. For instance, in 1879 in Kansas, an iron bridge was swept from its very foundation. On 18 March 1925, the "tri-state" tornado, covering Missouri, Illinois and Indiana, with a track of 219 miles, killed 689 and injured 2,027 people, and caused 17 million dollars property damage in 3.5 hours. In 1931, in Minnesota, a twister snatched up five train coaches, each weighing 70 tons, and dropped then 30 meters away. In 1955, a tornado roared through Illinois, picked up a car with two people in it, and transported it 30 meters away. On 3 and 4 April 1974, during a "super outbreak" of 148 tornadoes across 13 states, more than 300 people were killed. The largest number of tornadoes has been observed in Oklahoma. During a period from 1953 to 1976,1326 instances of tornadoes were recorded in this state.

Other atmospheric vortices. Waterspouts and dust devils are tornado look-alike phenomena, with much weaker intensity. Waterspouts consist of columns of rotating winds and are formed over warm waters. Dust devils are small, rapidly rotating winds over hot ground, made visible by dust and swept up debris. Dust devils develop over land on clear, dry, hot afternoons.

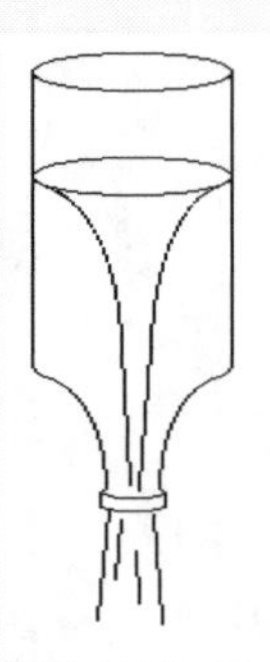

8.6. TORNADO-LIKE VORTICES (1)

Materials: bottle, water.

Procedure: Fill a bottle about 3/4 full of water. Rapidly swirl it, invert and let it drain. A tornado-like whirlpool will form in the bottle. The vortex persists until the water has drained.

Comment: When water is swirled in a bottle which is not draining, the vortex is stopped shortly by friction. However, when water is swirled in a bottle which is draining, the rotation is prolonged. The decrease of the amount of water in the bottle increases its spin, and counteracts against friction.

8.7. TORNADO-LIKE VORTICES (2)

Materials: jar with a cap, detergent, water.

Procedure: Fill the jar about two-thirds full of water. Add a small amount (half of a teaspoon) of laundry detergent. Rapidly swirl the liquid in the jar. Observe that a small whirlpool forms.

Speed of tornado winds. The 19th century American meteorologist Elias Loomis (1811-1889) observed that the violent winds accompanying tornadoes can strip large birds (such as fowl, geese, hens or turkeys) of their feathers. From this fact, he attempted to assess the speed of tornado winds. He used a small canon and substituted a chicken carcass for a cannon ball. The gun was pointed vertically and fired. Loomis reported that the chicken's body reached a velocity of about 340 mi/h and was torn into small pieces, only part of which could be found. Nevertheless, the feathers were pulled out clean. From this evidence, Loomis concluded that the speed of tornado winds must be below 340 mi/h, because tornadoes pulled out feathers without mutilating birds' bodies.

CONVECTION IN THE ATMOSPHERE. The vertical motion of fluid is called *convection*. Convection occurs, for instance, when a shallow layer of fluid is heated from below and cooled from above. The result is a strikingly uniform pattern of convection cells. This phenomenon was first reported in 1900 by Henri Bénard (1874-1939).

Bénard cells are easily visible in a cup of coffee. When hot coffee is poured into a cup, and cold cream is immediately added, the cell structure is clearly observable at the surface of the liquid. The first theoretical studies of this type of convection were made by Lord Rayleigh in 1916.

Convection in a cup of coffee

8.8. BENARD CELLS

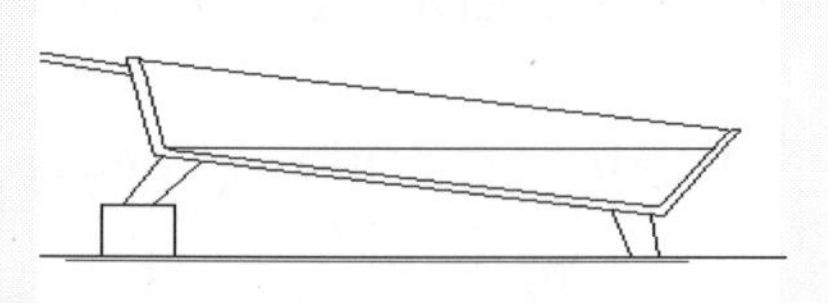

Materials: frying pan, corn oil, cocoa, stove burner.

Procedure: Pour corn oil in a clean frying pan so that there is a layer of oil about 2 cm deep. Heat the tilted pan. To visualize the flow, drop in a little cocoa. Oil will break into a complicated network of convective circulation known as Benard cells. The movement of individual particles of powder may be seen as rising near the center of a cell and falling near the sides.

Comment: Convection in this experiment is dependent on the temperature difference between the top and the bottom and the depth of the oil. The role of the oil's depth is clearly demonstrated when the frying pan is tilted. No cells are visible until the depth reaches a certain "critical" value. At depths larger than "critical", the cell size increases proportionally to the depth. Each hexagonal cell is made up of six equilateral triangles.

Convection in the atmosphere

In nature, convection occurs frequently, but usually is invisible to the naked eye. Convective patterns in the atmosphere (hexagonal cells and horizontal rolls) can often be seen from high flying aircraft or a satellite perspective, if clouds are present. The accompanying photo depicts characteristic cloud bands which indicate the presence of rolls in the atmosphere. In 1938, similar patterns were identified on lake surfaces by Irving Langmuir (1881-1957). The photograph on the following page shows foam bands due to horizontal roll vortices in water.

Convective patterns on a lake

Convection can be generated by the temperature contrasts at the Earth's surface. One example is the difference of temperature between the sea and the land. During the daytime, the coast heats more rapidly than the sea. Conversely, at night, land cools more quickly than the sea. The resulting wind, occurring along coasts and caused by thermal forcing, is called a *land/sea breeze.* The land breeze blows toward the sea at night, and the sea breeze toward the land during the day.

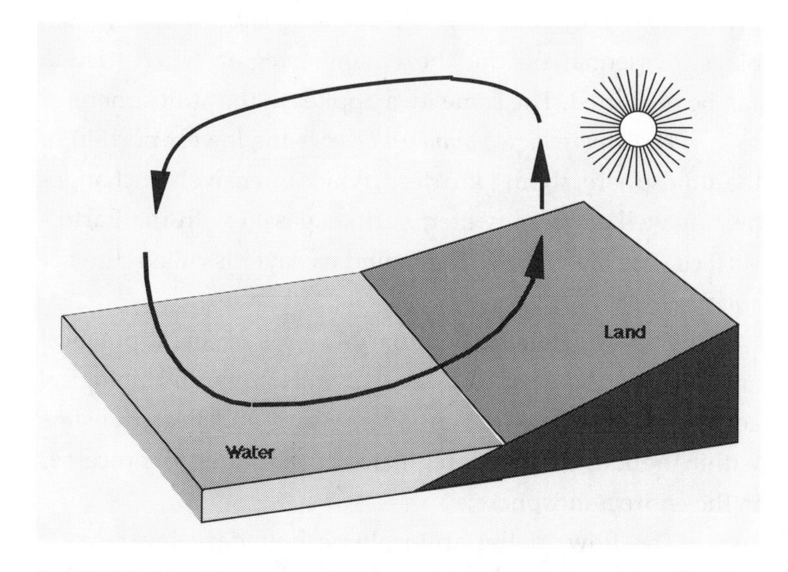

Sea breeze

8.9. BREEZE-LIKE CONVECTION

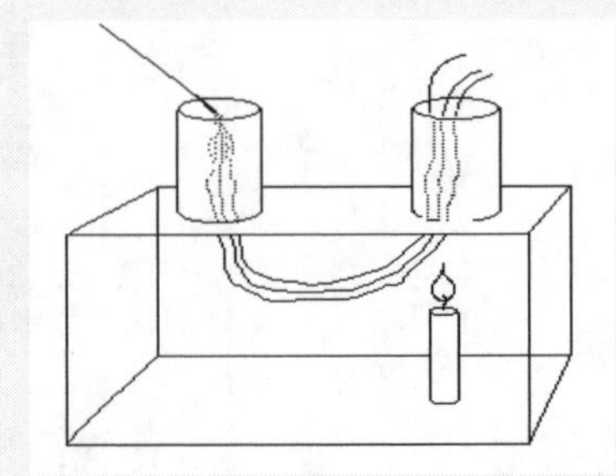

Materials: box with a glass front, two paper pipes, candle, incense stick, matches.

Procedure: Make a box with a glass front and two paper pipes as chimneys. Light the candle and the incense stick. Place the candle below one chimney and hold the incense stick above the other. The smoke will descend through one chimney and rise in the other, indicating the air circulation.

Explanation: Heat generated by the candle causes a breeze-like circulation.

ATMOSPHERIC BOUNDARY LAYER. In 1904, Ludwig Prandtl (1875-1953) introduced the idea that a flow about a solid body can be divided into two regions: a thin layer in proximity to the body, called the *boundary layer*, where friction plays an essential role, and the remaining region, where friction can be neglected. The same idea applies to the atmosphere.

The *atmospheric boundary layer* is the lowest portion of the atmosphere, about 1 km deep, which intensively exchanges heat as well as mass (water, various gases) with the Earth's surface. The layer above the boundary layer is called the *free atmosphere*.

The atmospheric boundary layer is of great practical and scientific importance. Essentially, all human and biological activities take place in this layer. Water and heat transfers within the boundary layer regulate a broad variety of processes in the entire atmosphere.

The flow in the atmospheric boundary layer has a

turbulent character. Turbulence is an essential part of the mechanism which disperses air pollutants resulting from anthropogenic activities. Turbulence is also crucial for the efficiency of many natural processes, such as the evaporation of water, dissipation of fog, and dispersion of plant seeds. It also contributes to the structural fatigue of steel and concrete constructions (e.g., on 8 November 1940, a large "Tacoma Narrows Bridge" was destroyed by wind generated vibrations).

Wind velocity in the boundary layer decreases to zero near the surface due to friction. The wind direction varies with height because of the influence of the Coriolis force. The latter fact was discovered by Fridtjøf Nansen (1861-1930) and Valfrid Vagn Ekman (1874-1954).

Fridtjøf Nansen was a famous Norwegian explorer of the Arctic, and the recipient of the 1922 Nobel Peace Prize. By examining data from the 1893-1896 Norwegian North Polar Expedition, Nansen noted that sea ice did not drift in the direction of the wind, but at an angle of about 40° to the right of the wind direction. Puzzled by this observation, Nansen discussed the problem with Ekman. They both concluded that each layer of the sea must be set in motion by the layer just above it, and successively more deflected by the balance of the Coriolis and friction forces. Ekman explained this phenomenon mathematically in two papers. The first one was published in 1902. The second one, *On influence of the Earth's rotation on ocean currents*, was published in 1905. In honor of Ekman's work, the spiraling of the currents in the ocean is named the *Ekman spiral.*

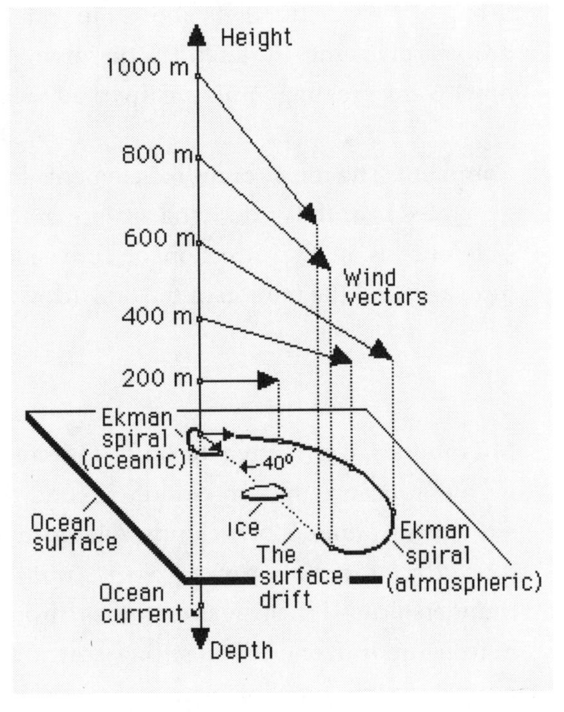

Ekman spirals on the Northern Hemisphere

The *Ekman spiral* is also observed in the atmospheric boundary layer (see the figure on the previous page), where winds spiral 5°-45° with height, clockwise in the Northern Hemisphere, but counterclockwise on the Southern Hemisphere. The spiraling effect in the atmosphere is caused by the balance of the Coriolis, pressure gradient, and friction forces.

8.10. "INVERSE THERMALS" IN A TANK

Materials: glass tank, food coloring, salt, water, eye dropper.

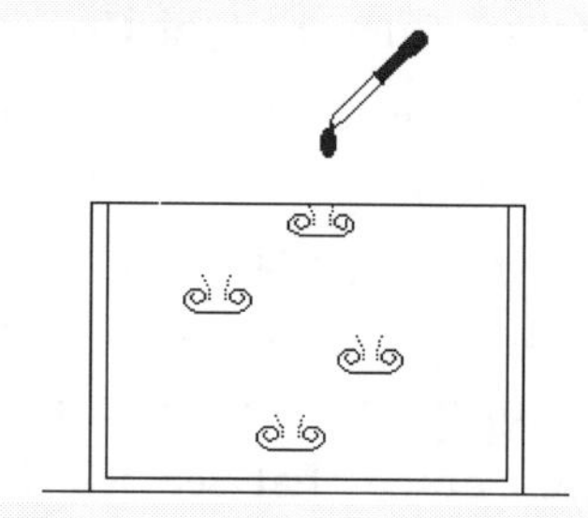

Procedure: Fill the tank with water. Fill the eye dropper with the solution of food coloring, water, and salt. Holding the eye dropper a few centimeters above the water surface, release some droplets of ink into the tank. The dissolving ink droplets will bulge downward creating multiple inverted mushroom-like elements.

Comment: The mushroom-like elements look similar to thermals generated near the underlying surface of the boundary layer. The difference is in the direction of convection in the atmosphere (upward) and the motion in the tank (downward).

Microbursts. Convective updrafts and downdrafts in the boundary layer are normally near 1 m/s in cloudless conditions. They are about 10-50 times stronger in thunderstorm clouds. A large number of airplane crashes have been caused by the formation of *microbursts,* updrafts and downdrafts beneath severe thunderstorms. When an aircraft flies through a microburst, it first encounters a strong updraft and then a strong downdraft. If the plane is near the ground, often the pilot cannot recover, and a crash is inevitable.

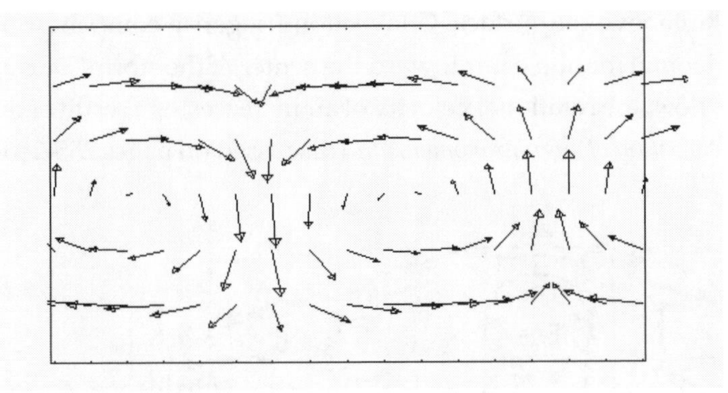

Computer generated display of a convective flow over a heated surface.

HURRICANES. A *hurricane* (also known as a *typhoon* in the western North Pacific) is another example of an extremely intense, long-lived vortex in the atmosphere. Its name comes from *Huracan*, a Carib god of evil. Hurricanes are migratory tropical cyclones, initiated by moist convection in a warm and steamy tropical marine boundary layer. Mature hurricanes extend over a distance of 600-1500 km, produce winds exceeding 120 km/h, and are accompanied by extremely heavy rains. Air pressure in their centers can drop even below 900 mb. Extensive damage associated with hurricanes is caused by torrential rains, strong winds, and the flooding of coastal regions.

Hurricane formation is possible when the temperature of the top layer of the ocean is higher than 26.5°C. Consequently, the hurricane season lasts from late May to late November. The contribution of the Coriolis effect (which vanishes at the Equator) in maintaining vortex motion causes hurricanes to form over the tropical oceans at latitudes of 5° - 20° (except the South Atlantic and the eastern South Pacific).

A hurricane can be compared to a heat engine, which transforms the heat of the tropical ocean into air motion. The energy produced by a hurricane in only one day is equivalent to the annual electric energy production in the United States. The engine cycle consists of a few stages. First, the heat energy at the ocean's surface is transferred into a latent heat of evaporation. This energy is lifted by updrafts in a process of thermal convection. The latent heat is liberated by water vapor condensation. As a result, convection intensifies, and cloud tops are even able to reach altitudes of about

15 km above the surface. Convection triggers a compensating horizontal motion of air toward the center of the storm. During this flow, the rotational velocity of air increases, as a result of the *conservation of angular momentum* (discussed on pages 255-256).

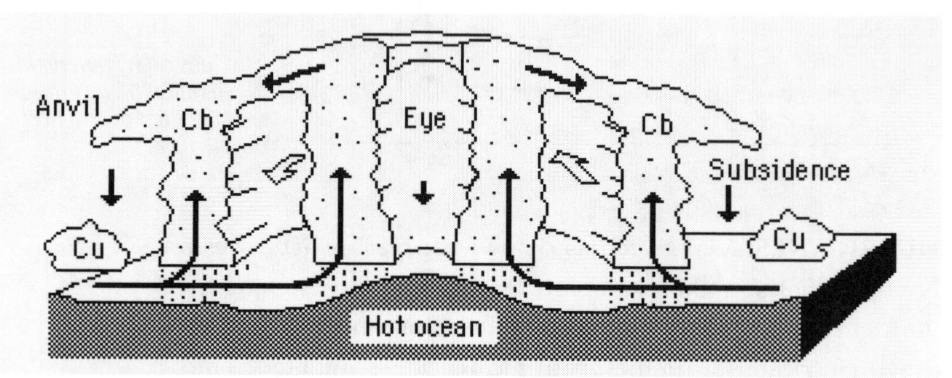

Idealized structure of a hurricane

The very center of the hurricane is occupied by a cloud-free *eye*, with slowly subsiding air. The eye is surrounded by an intense cumulonimbus development called an *eye-wall*. The higher pressure over of the storm center pushes air outward. Anvil-shaped cloud tops merge and form a spiralling dome. Low air pressure in a hurricane center causes the sea level rise (about 0.5 m for every 50 mb pressure drop). Resulting *storm surges* are driven by wind over low-lying coastal areas causing floods.

Hurricane names. Since 1979, names have been assigned to North Atlantic and Eastern Pacific hurricanes, based on two special lists. The lists change each year and consist of male names alternating with female names. For instance, the list prepared for North Atlantic hurricanes in 1996 includes the following names: Arthur, Bertha, Cesar, Diana, Edouard, Fran, Gustav, Hortense, Isidore, Josephine, Klaus, Lili, Marco, Nana, Omar, Paloma, Rene, Sally, Teddy, Vicky, and Wilfred.

A hurricane begins as a local storm, called a *tropical disturbance*. Gradually, it grows into a *tropical depression*, a stronger storm with winds below 61 km/h. When winds are in the range between 61 km/h - 115 km/h, the cyclone is called a *tropical storm.* Only a few tropical disturbances, depressions, or storms develop into full-fledged hurricanes. When a hurricane moves over land or colder water, it loses its energy, and eventually dies.

Hurricane flights. Col. Joseph P. Duckworth (pilot) and Lt. Ralph O'Hair (navigator) were the first aviators to enter the eye of a hurricane. On 27 July 1943, they flew a light, single engine AT-6 over a hurricane extending over the Gulf of Mexico and Texas. The flight initiated regular reconnaissance flights during hurricane seasons, and led to the improvement of hurricane tracking and forecasting abilities.

CAN WEATHER BE PREDICTED? For centuries, people have tried to forecast the weather. They carefully observed signs of nature in the sky and on the Earth, looking for hints which would make such predictions possible. Everything, from the color of the sky to the behavior of plants and animals, was thought to provide important clues about weather. For instance, in about the 6th century B.C., The Chinese philosopher Lao-tsu, in his book *Tao-te-ching* maintained that "*a strong wind does not last the whole morning, a heavy rain does not last the whole day*" (this was probably also an allegory designed to explain human behavior). The Greek philosopher, Aratos of Soli (3rd century B.C.), stated that "*clear weather occurs if a crescent moon appears sharp and clear. If it is surrounded by a reddish aureole, wind will prevail, but if both its ends appear blunt, a heavy rain will follow.*" The 16th chapter of the Gospel according to St. Matthew reads: "*when it is evening, it will be fair weather, for the sky is red. And in the morning, it will be foul weather today: if the sky is red and lowering*".

There have been many folk adages on this subject: "*Evening red and morning gray are sure signs of pleasant day*", "*Trout jump high when a rain is nigh*", "*Sound travelling far and wide, a stormy day will betide*", "*Birds flying low, expect rain and a blow*". There were many observations relating plant and animal behavior to weather. Scales on pine cones were observed to open during dry weather, and to close when rain was on the way. In fine weather, seaweed shriveled and seemed dry to the touch. When the rains came, the seaweed swelled and felt damp. Spiders were seen making alterations to their webs to suit the weather. For example, when rain threatened, spiders shortened their web filaments. On the contrary, seeing spiders running out slender filaments signaled fine weather.

Weather predictions were sometimes successful either by chance or method. For instance, in 1502, Columbus believed he had predicted the approach of a hurricane on the island of Haiti, by using local weather signs. On 9 December 1660, Otto von Guericke used a water barometer he had constructed to successfully predict a storm.

A reputable forecaster. A noted weatherman for one of the local radio stations stopped for a weekend to rest at a hotel in the mountains. When he was ready to go for a hike, the receptionist told him to stay indoors because it was going to rain. The weatherman disdained the advice and was drenched by a heavy shower. Astonished, he returned to the hotel to find out how the receptionist was able to predict the weather with such precision. When asked, the receptionist replied: "*That fellow who presents the weekend weather forecasts on the radio is such a notorious liar that whenever he promises fine weather, you can be sure it will rain!*"

Weather lore does not have scientific validity, and it often fails. When it happens, complaints multiply. "*Of all the silly, irritating foolishness by which we are plagued, this weather forecast fraud is about the most aggravating*" wrote Jerome K. Jerome in his *Three Men in a Boat*, "*it forecasts precisely what happened yesterday or the day before, and precisely the opposite of what is going to happen today*" Mark Twain wrote: "*Everybody talks about the weather, but nobody does anything about it*". So, can anything be done about it? Can weather be predicted?

The French mathematician Pierre Simon de Laplace (1749-1827) reasoned that the laws of nature (and weather) imply strict determinism and complete predictability. In 1776, in his essay *Philosophique sur les Probabilités,* he wrote: "*The present state of the system is a consequence of what it was in the preceding moment, and if we conceive of an intelligence which at a given instant comprehends all the relations of the entities of this universe, it could state the perspective positions, motions, and general effects of all these entities at any time in the past or future.*"

A different point of view was presented by Jules Henri Poincaré (1845-1912), who argued that arbitrary small uncertainties in the state of a system may be amplified in time and prevent correct predictions of the distant future. In 1908, in his *Science et Méthode,* he wrote: "*if we know exactly the laws of nature and the situation of the universe at the initial moment, we could predict exactly the situation of that same universe at a succeeding moment. But even if it were the case that the natural laws had no longer any secret for us, we could still only know the initial situation approximately. It may happen that small differences in the initial conditions produce very great ones in the final phenomena. A small error in the former will produce an enormous error in the latter. Prediction becomes impossible, and we have the fortuitous phenomenon.*"

Edward Lorenz (1917-), a professor at the Massachusetts Institute of Technology, was the first meteorologist who evaluated the validity of Poincaré's ideas. In 1961, Lorenz ran a simple numerical model of atmospheric convection on a Royal McBee computer at the Massachusetts Institute of Technology in Cambridge. One day, Lorenz compared two simulations of his model. The first simulation was run for the full period of a forecast. In the second simulation, the period of the forecast was the same; however, the calculation was split into two runs. The first "half" run was calculated for a short time. Then, its results were used as an initial condition for the second "half" run, which was continued for the rest of the time period.

The results of both simulations, the complete one and the one which consisted of two parts, were expected to be the same. Lorenz was very surprised to see that they were quite different. Because the equation and coefficients were the same, he first thought that something went wrong with the computer. After additional investigation, he found the reason for this difference. The problem was caused by the accuracy of the initial condition in the second "half" run. The results of the first "half" run were stored in the computer's memory as numbers with six decimal places. To save time, Lorenz used only three decimal numbers for the initial conditions for the second "half" run. For instance, instead of typing 0.452385, he rounded the number and typed in 0.452.

"Lorenz's butterfly"

Lorenz carefully analyzed the results of the second portions of both simulations. He noticed that the results of both simulations in each grid point were initially very close, but with time, both solutions were farther and farther apart. Consequently, he concluded that only the short-term (a few hours to a few days) evolution of weather systems could be accurately predicted. Long-term weather behavior (months or years ahead), however, is unpredictable because equations describing the state of the atmosphere are extremely sensitive to the initial conditions. This phenomenon is called "the butterfly effect", since one might believe that even a butterfly can change the initial weather condition at a certain point. Thus, long-term predictability would be possible only if the initial conditions in the atmosphere were known with infinite precision.

In 1961, Lorenz published his conclusions in a paper *Deterministic Non-periodic Flow* in the *Journal of Atmospheric Sciences*. His paper went unnoticed for the next decade. But later, it triggered intense research on *chaotic systems.*

NUMERICAL WEATHER FORECASTING. The first step toward mathematical weather forecasting was done by Lewis Fry Richardson (1881-1953). Richardson believed that it might be possible to solve the complex equations of atmospheric motion by working them out in step-by-step computations. He described his ideas in his book *Weather Prediction by Numerical*

Process. The first draft of it was prepared in May 1916, and revised in 1916-1918, during World War I. During the battle of Champagne, in April 1917, the working copy was sent to the rear lines and was lost. Fortunately, it was recovered a few months later under a heap of coal. Finally, the manuscript was printed in 1922.

To put his idea into practice, Richardson required data from about 2,000 permanent weather stations collecting both surface and upper-air measurements around the Earth. He also envisioned the globe divided up as a checkerboard into rows and columns. He thought that 32 individuals could compute a forecast at one column to keep pace with the weather. If the column spacings were 200 km, he believed that 2000 active columns would suffice to complete a forecast on the globe. Consequently, 32 x 2000 = 64,000 calculators would be needed to predict the weather for the entire globe. Such an amount of computations would require a "forecast factory".

Richardson envisioned the "forecast factory" as a large hall in a theater (see the picture below), with the circles and galleries going around through the space usually occupied by the stage.

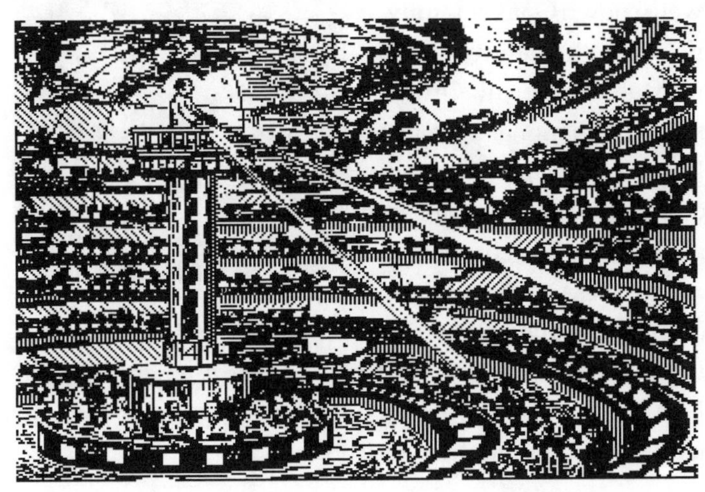

Richardson's "weather factory"

The walls of the hall were to be painted as a map of the Earth. The ceiling represented the northern regions, the tropics were in the upper circle, and the Antarctica in the pit. Thousands of mechanical calculators were to be used to prepare the weather forecast. Each calculator was planned to solve only one equation or part of it. Numerous light signs displayed the instantaneous solutions. Each number was displayed in three adjacent zones to maintain communication. The work of each zone was coordinated by an official of higher rank. From the pit a tall pillar was raised to half the height of the hall. On top of this pillar, the man in charge of the hall would sit. He was surrounded by a team of assistants and messengers. His official duty was to maintain a uniform speed of computations in all parts of the hall. For this purpose, he used a beam of red light to point any group of operators which was running ahead, and a beam of blue light to signal operators, who were behind with their calculations.

Lewis Fry Richardson (1881-1953)

Richardson was born in Newcastle-on-Tyne, Great Britain, to a well-known Quaker family. He obtained his Ph.D. in mathematics in 1916. Before joining the Meteorological Office, Richardson worked at the National Physical Laboratory, and also at a tungsten lamp factory. During World War I, he was a volunteer ambulance driver. When the Meteorological Office was taken over by the Air Ministry, for pacifist reasons, Richardson turned away from meteorology. He pioneered a completely new subject, a mathematical and psychological analysis of the causes for war, which he described in terms of ordinary linear differential equations. He earned a B.Sc. in psychology in 1919.

In order to demonstrate the validity of his method, Richardson made a forecast at two grid points. He used five *primitive* equations: hydrostatic, continuity, thermal-energy, and two horizontal-momentum equations. To solve these equations, he replaced derivatives by finite differences: $dF/dt = [F(t+\Delta t) - F(t-\Delta t)]/(2\,\Delta t)$, where t is time, and Δt is the time increment.

As a result of his work, Richardson laid the foundation for solving differential equations by adopting *finite-difference* equations, and by devising a numerical method of solution. The date chosen by Richardson for the initial data was 20 May 1910. He used weather maps, prepared by Bjerknes, during his tenure at the University of Leipzig.

The obtained results were "catastrophic", a 145 millibar increase of pressure in six hours. Richardson believed that this failure resulted from inadequacies of upper wind data. He did not realize that the six-hour computational time interval which he used was too long, and caused accumulation of numerical errors.

Richardson's dream of the numerical weather prediction came true in the mid 1940s, with the construction of the ENIAC (Electronic Numerical Integrator and Computer), the first electronic computer. The ENIAC, constructed by two Americans, John W. Mauchly (1907-1980) and John P. Eckart, Jr. (1919-), was an enormous device with 18,000 vacuum tubes, weighing 30 tons, and filling 150 m^2 of space. A group of meteorologists, upon hearing about this invention, suggested using computers in weather forecasting. John von Neumann (1903-1957) and his colleagues at Prinston University began working on the problem. The first test was made in March 1950. It was based on the mathematical model designed by Jule Charney (1917-1981). It produced the first successful numerical weather forecast after about 24 hours of calculations.

Today, computers are much speedier. For instance, the American made "CRAY Y-MP C90" can perform 16 billion operations per second. A modern 24-hour forecast requires about 10 billion arithmetical operations, and can be obtained in a very short time.

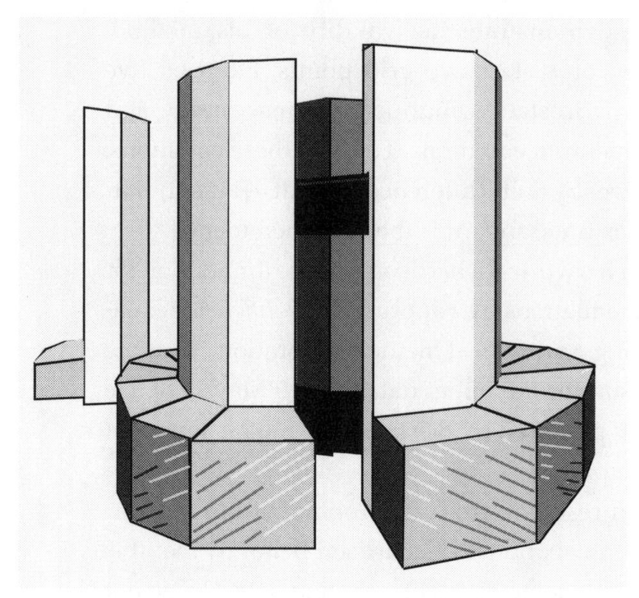

Supercomputer CRAY Y-MP

8.11. NUMERICAL FORECAST

Materials: calculator or personal computer.

Procedure: Imagine that the variation in time of a meteorological parameter X can be described in terms of the following prognostic equation: $X_{k+1} = 3.8 * X_k * [1 - X_k]$, where k = 0, 1, 2, 3,..., T, are instants of time counted in hours, X_{k+1} is the value of the parameter in the next (k+1) moment of time, while X_k is the value of the parameter in the previous (k) moment of time. For example, when k = 0 the prognostic equation reads: $X_1 = 3.8 * X_0 * [1 - X_0]$, for k = 1 the prognostic equation is: $X_2 = 3.8 * X_1 * [1 - X_1]$, e.t.c. Assume that the initial value of the parameter is $X_0 = 0.5$. Find the value of X_T, for T = 20 hours (if you have a calculator) or for T= 200 hours (if you have a computer). Plot all T values of the parameter X using Cartesian coordinates with the horizontal axis representing time and the vertical axis representing the parameter X.

MODERN FORECASTING TOOLS. Modern meteorology uses a number of sophisticated operational tools, such as radars and satellites. The first meteorological radar was built in 1942. The first meteorological satellite was launched in 1960.

The word radar is a contraction for "radio detection and ranging". Radar is an electronic instrument which senses the atmosphere by means of electromagnetic radiation. *Weather radar* sends out beams of very short impulses of radio waves. Objects in the atmosphere, such as rain drops, snow crystals, hail stones, insects, and dust particles reflect some of the radio waves back to the radar antenna. The returning signals are electronically converted into pictures showing the location, type, intensity of clouds, and precipitation.

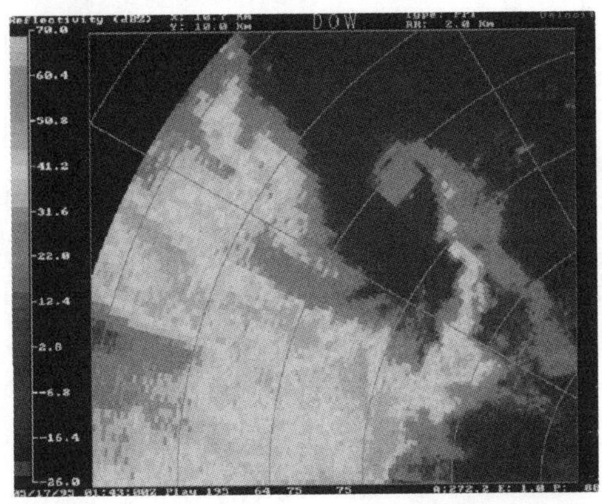

A Doppler radar echo of a tornado

Some weather radars can detect not only reflectivity, but also the frequency change in returning radio waves. These radars are called *Doppler radars*. The operation of Doppler radars is based on the fact that the reflected radar waves decrease their frequency when reflecting objects are moving away from the radar's antenna, and increase their frequency when the reflecting objects are moving toward it. The frequency shift can be electronically converted into useful information on wind velocities.

The Doppler effect. The Doppler effect refers to the principle discovered in 1842 by the Austrian scientist, Christian Doppler (1803-1853). The principle relates the pitch (frequency) of sound to the relative motion of the sound source with respect to an observer. Even though Doppler worked out his ideas for sound waves, the same idea applies to radar's radio waves. Doppler's principle was first tested during an unusual experiment in 1842, in Holland. A musician was placed on a flat car and asked to play notes on a trumpet. The flat car was pulled by a locomotive back and forth, at different speeds. On the ground, observers with a sense of absolute pitch recorded the notes as the train was approaching and receding.

Presently, radar data is provided by more than 130 radar sites around the United States. This data can be combined and viewed in single images on national and regional scales.

Satellites (which means "attendants" or "guards" in Latin) provide a useful tool for locating and tracking weather systems around the globe. The first weather satellite was Tiros, launched by the U.S. in April of 1960. Satellite cameras are sensitive to visible and infrared radiation and enable continuous monitoring of cloud patterns both during the day and during the night. The coldest (high) clouds are displayed in bright, white shades, while the warmest (low) clouds appear dark gray in color. Cloud patterns are analysed based on images from both, geostationary and polar orbiting satellites. Geostationary satellites operate at an altitude of approximately 22,000 miles (36,700 km), over fixed areas of the Earth. These satellites include GOES-7 (Western U.S., Eastern Pacific Ocean), GOES-8 (Eastern U.S., Central and South America, Atlantic Ocean), METEOSAT-5 (Europe, Africa, Middle East) and GMSSAT (Pacific Ocean). Satellite imagery for the remaining areas are provided by polar orbiting satellites, which orbit the Earth 14 times a day, at an altitude of approximately 520 miles (870 km). Due to the Earth's rotation, polar orbiting satellites are able to view each spot on the globe twice a day.

Polar satellite imagery of Hurricane Erin approaching Florida on 1 August 1995.

EPILOGUE

It is vain
to do more
what can be done
with fewer.
(William of Occam)

Throughout the ages, observations of the atmospheric phenomena have triggered interest in the search for answers to elusive and puzzling meteorological questions. As the scientists of past generations took their places in the annals of history, distinct theories on the atmosphere were formulated, and then examined through experimentation. The purpose of these scientific investigations was to improve the understanding of weather, and to make the knowledge of its processes useful, clear and factual. The triad -- history, theory and experimentation -- became interlaced in the delivery of scientific truths -- for all who have been interested in theoretical as well as popular approaches to meteorology. This layering effect took the study of meteorology to grand heights, where it continues presently in its broadest range of applications.

Nowadays, engineers, architects, builders, astronauts, pilots, sailors, biologists, farmers, geologists and marine scientists have to be aware of meteorological concepts. All human activities take part in the ocean of the atmospheric air. The atmosphere provides us with oxygen for breathing, and ozone as a protection against the ultraviolet solar radiation. The atmosphere is also the parent and sustainer of all vegetable life. Strong winds and their accompanying gusts have a variety of effects on skyscrapers, towers, bridges, chimneys, power transmission lines, etc.

Waste gases produced by industrial processes are released in the atmosphere for disposal, and their dispersal is dependent on the state of the atmosphere. Aviation and marine activities are strongly dependent on weather conditions. Aircraft are subjected to various weather hazards during take-offs, actual flights, and also during landings. These hazards include flight conditions caused by turbulence, icing, or thunderstorms.

The broad range of meteorological applications includes weather prediction, as well as various problems related to aviation, sea and land transport, air pollution modeling. Perhaps the most important applications of meteorology are those related to weather prediction. Weather forecasts are routinely used in civil and military activities, and also in planning tracks for trade and cruise ships, activities for fishing boats, as well as for launches of space shuttles.

The science of meteorology has grown through the centuries. The latest developments were triggered by the progress in electronics, applied mathematics, and in introduction of various measurement techniques. The most modern technology available applies sodars, lidars, radars, meteorological satellites, instrumented airplanes, and also the largest and the fastest computers, capable of adding billions of numbers in seconds. Past and present accomplishments, as well as new challenges, allows meteorology to be considered, without a doubt, one of the most interesting branches of modern science.

METEOROLOGICAL TIME-TABLE

500 B.C.: Parmenides classifies world climates,

ca. 350 B.C.: Aristotle writes "Meteorologica",

330 B.C.: Hippocrates writes a treatise on climate and medicine, "Air, Waters, and Places",

300 B.C.: Theophrastus writes a meteorological treatise, "On the Signs of Rain, Winds, Storms and Fair Weather",

300 B.C.: Eratosthenes evaluates the radius of the Earth; Euclid formulates the law of reflection,

62 A.D.: Heron writes "Pneumatica",

1025: Alhasan evaluates the height of the atmosphere,

1170: Aristotle's "Meteorologica" is translated into Latin in Italy,

1202: Arabic numbers are introduced to Europe,

1450: Cardinal Cusa invents the balance hygrometer,

1593: Galileo invents thermoscope; Napier invents logarithms,

1620: Snell formulates the law of refraction,

1626: Santorio invents the cat gut hygrometer,

1637: Descartes invents analytical geometry,

1638: Galileo determines that air has weight,

1644: Torricelli invents the barometer,

1648: Pascal determines that air pressure decreases with height,

1654: Von Guericke evaluates the density of air,

1657: Von Guericke performs the Magdeburg experiment; Accademia del Cimento is founded in Florence,

1659: Huygens discovers centrifugal and centripetal forces,

1661: Boyle and Hooke discover the first pressure law,

1675: Leibnitz introduces derivatives,

1680: Papin discovers that the boiling point temperature decreases with decrease in pressure,

1687: Newton publishes "Principia",

1738: Daniel Bernoulli publishes "Hydrodynamica",

1742: Celsius proposes the thermometric scale,

1751: Le Roy defines " dew point" and "precipitation",

1753: Black discovers carbon dioxide,

1755: Cullen invents the wet and dry bulb thermometer, Franklin collects thunderstorm electric charges, Euler formulates equations of inviscid fluid motion,

1764: Black makes distinction between temperature and heat,

1766: Cavendish discovers hydrogen,

1770: Rutherford discovers nitrogen,

1774: Scheele and Priestley discover oxygen, Lavoisier lays the foundation of modern chemistry,

1783: First manned balloon flight is performed by the Montgolfier brothers; the first scientific balloon flight is performed by Charles and Roberts;
de Saussure invents the hair hygrometer,

1800: Herschel discovers infrared radiation,

1801: Ritter discovers ultraviolet radiation,

1802: Gay-Lussac publishes the gas law; Dalton defines "relative humidity",

1803: Howard presents his cloud classification,

1811: Avogadro explains gas molecules,

1823: Laplace derives the hydrostatic equation,

1824: Carnot presents the general theory of heat engines,

1826: Brandes prepares the first pressure chart; the first national weather services established in Belgium,

1827: Navier formulates equations of viscous fluid motion,

1835: Coriolis discovers the laws of bodies in motion on a spinning surface,

1851: Lord Kelvin formulates the main laws of thermodynamics,

1855: Ferrel explains the general circulation of the atmosphere,

1875: Coulier discovers the role of condensation nuclei,

1883: Reynolds describes laminar and turbulent flows,

1888: Abercromby classifies pressure systems,

1892: Lord Rayleigh and Ramsay

discover argon,

1898: Ramsay discovers helium,

1899-1902: De Bort performs the first study of the upper atmosphere,

1904: V. Bjerknes lays the foundation of synoptic meteorology,

1905: Ekman explains flow in the boundary layer,

1920: Theory of atmospheric fronts is developed by V. Bjerknes, H. Solberg, and J. Bjerknes,

1922: Richardson lays the foundation of numerical weather prediction,

1928: Molchanov develops the first radiosonde,

1933: Bergeron initiates research on cloud formation,

1942: Weather radar developed,

1944: Jet-stream winds discovered,

1946: Invention of sodars; first rockets are used to investigate the atmosphere; first rain making experiments performed by Schaefer and Langmuir,

1950: World Meteorological Organization (WMO) established,

1950: The first numerical weather prediction is made on ENIAC computer,

1957: The first satellite is launched,

1960: The first meteorological satellite is TIROS-1 launched,

1963: Lorenz applies chaos theory to explain weather predictability,

1990: First Doppler-radar network is introduced in meteorological service (in the U.S.A.).

GLOSSARY

A

Absolute humidity-the ratio of the mass of water vapor present to the volume occupied by the mixture of water vapor and dry air.

Absorption-the process by which radiant energy incident on any substance is retained by that substance.

Adiabatic lapse rate-the rate (equal to 0.98°C per 100 m) of decrease of temperature with height in a parcel of unsaturated air which moves vertically.

Adiabatic process-the process performed in such a way that no heat is added to or removed from the system.

Advection-the horizontal transport of air carrying certain properties.

Aerosols-Various liquid or solid particles suspended in the atmosphere.

Air density- the ratio of air mass to volume.

Air mass-a large body of air in which temperature and humidity in a horizontal plane are uniform.

Altocumulus-layers or patches of medium level clouds often with a waved appearance, composed mostly of liquid water droplets, but also of supercooled water droplets or ice crystals.

Altostratus-layers of flat, thick gray veil, composed mostly of liquid water droplets, but also of super cooled water droplets or ice crystals.

Anemometer-an instrument for measuring wind speed.

Aneroid-a type of a barometer in which the volume of a sensor changes with pressure.

Anticyclone-a system of high atmospheric pressure, with the highest pressure in the center and with a clockwise circulation on the Northern Hemisphere and counter-clockwise circulation on the Southern Hemisphere.

Anvil cloud-name of the flattened upper portion of the cumulonimbus cloud.

Arctic air-a cold air mass developed over Arctic regions of the Earth.

Atmosphere-the body of air surrounding the Earth.

B

Barograph-a continuously recording barometer.

Barometer-an instrument for measuring air pressure.

Buoyancy-the upward force exerted on an object immersed in fluid.

C

Celsius scale-a temperature scale with 100 units between the freezing point (0°) and the boiling point (100°) of pure water at standard sea-level pressure.

Centrifugal force-the apparent force in a rotating system deflecting a rotating mass radially outward from the axis of rotation.

Centripetal force-the force exerted on a rotating mass radially toward the axis of rotation, and equal to the centrifugal force but in the opposite direction.

Cirrocumulus-a high cloud consisting of ice crystals and appearing as a thin sheet of small white puffs.

Cirrostratus-a high cloud consisting of ice crystals and appearing as a white veil, which sometimes produces a halo.

Cirrus-a high cloud consisting of ice crystals appearing as thin, white feather-like patches or narrow bands, made of ice crystals.

Climate-statistical features of weather over a specified region for the specified interval of time.

Coalescence-merging of two droplets of cloud.

Cold front-an edge of a cold mass of air which replaces warmer air.

Condensation-the change of gaseous state into a liquid state.

Condensation level-the height at which condensation begins in a rising parcel of air.

Condensation nuclei-small particles in the air upon which condensation begins.

Conduction- the transfer of heat by molecular action through contact.

Constant pressure chart-a weather chart at a constant pressure with contour lines indicating heights of the pressure surface above sea level.

Contour line-line on a weather chart which connects points of the same height.

Contrails-long streaks of cloud formed by high flying airplanes.

Convection-atmospheric motion which is predominantly vertical.

Convergence- the net inflow of air into a given area.

Coriolis force-a deflective force resulting from the Earth's rotation, which acts to the right of the wind direction in the Northern Hemisphere and to the left in the Southern Hemisphere.

Corona- a colored circle or arc around

the sun or moon, formed by diffraction of light by a veil of water droplets, with colors from blue inside to red outside.

Cumulonimbus-massive clouds of large vertical extent, capable of producing lightning, strong wind, hail, and tornado.

Cumulus- well defined, white clouds in the form of domes or towers.

Cyclone-a system of low atmospheric pressure, with the lowest pressure in the center, with a counterclockwise circulation on the Northern Hemisphere and clockwise circulation in the Southern Hemisphere, accompanied by fronts, and associated with stormy weather.

D

Density-the ratio of the mass of any substance to its volume.

Deposition-a direct change of state from vapor to solid.

Dew point-the temperature to which air has to be cooled in order to attain saturation of water vapor.

Diffraction-bending of light wave beyond opaque objects.

Diffusion-exchange of fluid properties due to chaotic motions of fluid molecules or small portions of fluid.

Divergence- the net outflow of air from a given area.

Doldrums-the equatorial zone of calm or light winds.

Downdraft- a small scale downward flow of air.

Drizzle-a form of precipitation with very small water drops.

Dry adiabatic lapse rate-see: adiabatic lapse rate.

Dry bulb temperature-a temperature registered by an ordinary thermometer.

E

Eddy-a local irregularity of wind caused by turbulence.

Ekman spiral-the curve showing the variation of wind in the lower atmosphere.

El Niño-Extreme patterns of weather caused by the periodic warming of the central and eastern equatorial Pacific.

Energy-an ability to do work. The basic forms of energy are: kinetic energy, potential energy, and heat. Energy of one joule causes force of one newton to act through a distance of one meter.

Entropy-an unavailability of energy for further energy transformations.

Equinox-a moment of time when night

and day last 12 hours equally.

Evaporation-change of state from liquid to gas.

F

Fahrenheit scale-a temperature scale with 32° as the freezing point and 212° as the boiling point of water at standard sea level pressure.

Force-an ability to accelerate or decelerate an object, measured as a product of mass and acceleration. Force of one newton causes the mass of one kilogram to change its velocity in one meter per second during one second.

Freezing-change of state from liquid to solid.

Front-a transition zone between two adjacent air masses.

Frontogenesis and Frontolysis-the initial formation and dissipation of a front.

Funnel cloud-a tornado cloud expanding downward from the parent cumulonimbus cloud, but not touching the ground.

G

Glory-colored rings around shadows of objects on clouds and fogs, as a result of refraction and diffraction.

Gradient-change in value of a parameter per unit distance.

Gravity-the force of attraction between any two material objects.

Gravitational acceleration-acceleration due to the force exerted by the Earth, equal to about 9.8 m/s^2.

H

Hail-a form of precipitation composed of solid balls of ice.

Halo-a colored or whitish, 22° or 46° (rare) circle or arc around the Sun or Moon, with coloration from red inside to blue outside, caused by the refraction of sunlight through a thin veil of ice crystals.

High-an area of high barometric pressure, an anticyclone.

Humidity-water vapor content in the air, expressed as absolute, relative, specific humidity, or mixing ratio.

Hurricane-a tropical, non-frontal cyclone with winds in excess of 120 km/h.

I

Ice crystals-a type of precipitation composed of slowly falling needles, columns or plates.

Infrared radiation-invisible electromagnetic radiation with wavelengths from 0.8 to 1000 μm, producing the sensation of heat.

Intertropical Convergence Zone (ITCZ)-the zone near the Equator, where the northeast trades of the northern hemisphere merge with the southeast trades of the southern hemisphere, and characterized by the presence of persistent cumulonimbus clouds and showers.

Inversion-an increase of air temperature with height.

Isobar-a line of constant barometric pressure.

Isoline (or isopleth)- a line of constant value of any meteorological variable.

Isotherm-a line of constant temperature.

J

Jet stream-a horizontal, narrow, fast stream of air near the tropopause.

K

Kelvin scale-a temperature scale with zero degrees at the temperature for which molecular motion ceases. A degree on the Kelvin scale is equal to a degree on the Celsius scale.

Kinetic energy-the energy of an object due to its motion.

Knot-unit of velocity equal to one nautical mile per hour, which is about 0.46 m/s.

L

Laminar flow-slow, smooth and regular flow.

Land breeze-a coastal breeze blowing from land to sea or lake.

Lapse rate-the rate of decrease of temperature with height,

Latent heat-the amount of heat absorbed or released during the process of phase change. Its basic forms are: latent heat of condensation (or evaporation), latent heat of fusion (freezing or melting), latent heat of sublimation (or deposition).

Lidar- (contraction for "light detection and ranging") an electronic instrument which senses the atmosphere by sending a narrow beam of light.

Light-visible electromagnetic radiation with wavelengths from 0.4 to 0.7 μm.

Lightning-any form of visible electrical discharge from a thunder storm.

Low-an area of low barometric pressure, a cyclone.

M

Melting--change of state from solid to liquid.

Mesosphere-a layer in the upper atmosphere above the stratosphere.

Mesopause-a transition layer in the upper atmosphere, at an altitude of approximately 90 km, between the mesosphere and the thermosphere.

Meteorology-the science of atmospheric processes.

Millibar-a pressure unit.

Mirage-a refraction phenomenon displacing images of objects from their true positions near the Earth's surface due to strong vertical temperature contrasts.

Mixing ratio-the ratio of the masses of the water vapor and of the dry air, present in a volume of moist air.

Moist-adiabatic lapse rate-the variable rate (about 0.6°C per 100 m) of temperature change with height in a parcel of saturated air which moves vertically.

Moisture-atmospheric water in any of its three states.

Momentum-the product of mass and velocity.

N

Nimbostratus-a gray, horizontally extended, low cloud, which can produce rain or snow.

O

Occlusion-a composite of warm and cold fronts.

Occluded front-a zone between two different masses of air formed by merging of cold and warm fronts.

Ozone- the three-atom form of oxygen (O_3).

Ozone layer-a layer of concentrated ozone in the stratosphere, 20 to 40 km above the Earth's surface, which strongly absorbs ultraviolet solar radiation.

P

Parcel of air- a small volume of air.

Polar front-the semi-permanent front separating air masses of tropical and polar origins.

Potential energy-the energy which is stored in a body by virtue of its position in the field of gravity.

Precipitation-any form of water particles falling from the atmosphere to the Earth's surface.

Pressure-the ratio of a force exerted on any surface to the area of the surface.

Pressure gradient-the rate of pressure change per unit distance at a fixed time.

Prevailing westerlies-the dominant west-to-east motion of the atmospheric air near the surface over middle latitudes on both hemispheres.

Pseudo-adiabatic lapse rate- see: moist adiabatic lapse rate.

Psychrometer-an instrument containing dry and wet bulb thermometers used for measuring atmospheric humidity.

R

Radar- (contraction for "radio detection and ranging") an electronic instrument which senses the atmosphere (precipitation, clouds) by means of electromagnetic radiation in radio wave frequencies.

Radiation-the transport of energy in the form of electromagnetic waves.

Radiosonde-a balloon-borne system of instruments for measuring the vertical distribution of pressure, temperature, and humidity in the atmosphere.

Rain-a form of liquid precipitation.

Rainbow-a luminous arc on the sky due to refraction and reflection of light by drops of rain water.

Rawinsonde-a balloon-borne system of instruments used to infer the vertical distribution of pressure, temperature and humidity in the atmosphere, with the capability for determination of winds aloft through the radio tracking.

Refraction-bending of a beam of light by variations in atmospheric density.

Relative humidity-the ratio of the actual amount of water vapor in the air to the amount in the saturation state at given temperature, usually expressed in percent.

Ridge-elongated area of higher atmospheric pressure associated with an anticyclone.

Riming-the formation of granular deposits of ice due to rapid freezing of supercooled fog or cloud water droplets.

S

Saturation-the condition when the actual water vapor amount is the maximum theoretically possible at the given temperature and pressure.

Scattering-the process in which small particles suspended in the atmosphere disperse the incident solar radiation in all directions.

Sea breeze-a coastal wind that blows from sea to land.

Sensible temperature-the empirical indicator of weather condition which takes into account temperature, humidity and wind speed.

Shower-rapid and intense precipitation from a cumulonimbus cloud.

Sleet-ice pellets or, casually, mixture of rain and snow.

Smog-a mixture of smoke and fog.

Snow-solid form of precipitation composed of white or translucent ice crystals of hexagonal form.

Snow pellets-precipitation consisting of white, opaque, ice particles.

Sodar- (contraction for "sound detection and ranging") an electronic instrument which senses the atmosphere by means of sound waves.

Solar radiation-the total electromagnetic radiation emitted by the Sun.

Solstice-the moment of the shortest or the longest daylight period on the hemisphere.

Sounding-an upper atmosphere observation, as from a radiosonde.

Supercooled water-water at temperature below 0°C.

Specific humidity-the ratio by weight of the amount of water vapor to the amount of moist (dry air and water vapor) air.

Squall line-a non-frontal line of active thunderstorms.

Stable atmosphere-a thermal state of the atmosphere which suppresses vertical motion of air parcels.

Stationary front-a transition zone between warm and cold air which does not move or moves slower than 2 m/s.

Stratocumulus-a low, horizontally expanded cloud with whitish patches or puffs.

Stratopause-the transition layer between the stratosphere and the mesosphere, approximately 50 km above the Earth's surface.

Stratosphere-the portion of the atmosphere, from about 10 km to about 50 km, characterized by low moisture content, and steady or increasing temperature.

Stratus-a low, horizontally expanding, thin, gray cloud layer.

Sublimation-a direct change of state from solid to vapor.

Supercooled water-liquid water existing at temperatures below the normal freezing point.

Synoptic chart-a chart depicting the distribution of meteorological conditions over a large area at a given time.

T

Temperature-the degree of hotness or coldness measured on some temperature scale by means of thermometers.

Theodolite-an instrument for measuring vertical and horizontal angles.

Thermometer-an instrument for measuring temperature.

Thermosphere-outermost layer of the atmosphere in which temperature increases with height.

Thunderstorm- a local storm produced by a cumulonimbus cloud, accompanied by lightning, thunder and showers.

Tornado-a violently rotating column of air associated with a severe thunderstorm.

Trade winds-prevailing easterly winds blowing from the subtropical high pressure zone toward the Equator.

Tropopause-a transition zone between the troposphere and stratosphere, approximately 11 km above the Earth's surface.

Troposphere-the lowest 10 km thick portion of the atmosphere, where most weather phenomena occur.

Trough-elongated area of relatively low pressure associated with a cyclone.

Turbulence-irregular, chaotic flow.

Twilight-an interval of incomplete darkness before sunrise or after sunset.

Typhoon-hurricane in the western North Pacific.

U

Ultraviolet radiation-invisible electromagnetic radiation with wave lengths from 0.001 to 0.4 μm.

Units-in science only the metric system is acceptable. The metric system uses the following units:
time: second (s), hour (h),
length: meter (m), 1 kilometer (km) =1000 m, 1 m = 100 centimeters (cm), 1 cm=10 millimeters (mm), 1 m =1,000,000 micrometers (μm),
area: square meter (m^2)
volume: cubic meter (m^3),
mass: 1 kilogram (kg),
temperature: kelvin (K), celsius (°C),
force: 1 newton =1 kg m s^{-2},
energy (heat): 1 joule = 1 kg m^2 s^{-2},
pressure: 1 pascal (Pa) =1 N m^{-2}.
Other units can be converted into the metric system as follows:
1 m = 39.37 inches, 1 inch = 2.54 cm, 1 kg = 2.2 pounds, 1 pound = 0.45kg, 1 gallon=0.0038 m^3 = 3.785 liters (l), 1 l = 0.001 m^3, 1 millibar = 1 hectopascal (hPa) = 100 Pa,
1 millimeter Hg =133.32 Pa.

Unstable atmosphere-a thermal state of the atmosphere which enables amplified vertical motion of air parcels.

V

Vector-any quantity that is specified by direction as well as magnitude.

Virga-falling water droplets or ice crystals which evaporate before reaching the Earth's surface.

Vorticity-rate of rotation in a fluid flow, typically around vertical axis, positive in counter-clockwise circulation.

W

Warm front-an edge of a warm mass of air.

Water vapor-gaseous phase of water.

Wavelength-a distance between two adjacent crests of a wave pattern.

Weather-the instantaneous state of the atmosphere.

Wet-bulb temperature-the temperature measured with a wet-bulb thermometer.

Wet-bulb thermometer-a thermometer with a muslin-covered bulb used for evaluation of relative humidity using evaporation.

Wind-horizontal flow of air.

Wind direction-direction from which the wind is blowing, with angles decreasing counterclockwise from the North.

Wind speed-horizontal distance which the air passes in a unit of time.

Wind vane-an instrument for measuring wind direction.

Work-the transference of energy defined as a product of the exerted force and a resulting displacement.

INDEX OF SUBJECTS

INDEX OF NAMES

Underlined type denotes bibliographical section.

INDEX OF EXPERIMENTS

MOISTURE

FORCES

MOTION

SELECTED BIBLIOGRAPHY

General Meteorology:

Ahrens, C. D. 1988: *Meteorology Today*. West Publishing Company.

Freier, G., 1989: *Weather Proverbs*. Fisher Books.

Huschke, R.E. (ed.), 1959: *Glossary of Meteorology*. Amer. Meteor. Soc.

List, R.I., 1971: *Smithsonian Meteorological Tables*. Smithsonian Institution Press.

Moran, J. M. and M. D. Morgan, 1995: *Essentials of Weather*. Prentice-Hall.

Riordan, P. and P.G. Bourgot, 1985: *World Weather Extremes*. Report ETL-0416. U.S. Army Engineer Topographic Laboratories, Ft. Belvoir., VA.

Schaefer, V. and J. A. Day, 1981: *A Field Guide to the Atmosphere*. Houghton Mifflin Company.

Wagener, R. L., 1994: *The Weather Sourcebook*. Globe Pequot Press.

Williams, J., 1992: *The Weather Book*. Vintage Books.

Westervelt, A. B and W. T. Westervelt, 1982: *American Antique Weather vanes*, The complete Illustrated Westervelt Catalog, 1883. Dover.

History:

Asimov, I, 1964: *Asimov's Biographical Encyclopedia of Science and Technology*, Doubleday & Company.

Bell, T.A, 1986: Man of Mathematics. Simon & Schuster.

Asimov, I, 1964: *Asimov's Chronology of Science and Discovery*. Harper Collins Publishers.

Bates, C. C. and J. F. Fuller, 1986: *Americas Weather Warriors*, 1814-1985. Texas A&M University Press.

Bronowski, J. (ed), 1960: *Doubleday Pictoral Library of Science*. Doubleday & Company.

Bronowski, J. (ed), 1966: *The Ascent of Man*. Little, Brown and Company.

Burke, J. 1995: *The Day the Universe Changed* J. Little, Brown and Company.

Burke, J. 1995: *Connections*. Little, Brown and Company.

Cohen, I. B., 1980: *Album of Science from Leonardo to Lavoisier*. Charles Scribner's Sons.

Fleming, J. R., 1990: *Meteorology in America, 1800-1870*. The Johns Hopkins University Press.

Fleming, T, 1972: *The Founding Fathers: Benjamin Franklin, a Biography in His Own Words*. Newsweek. Distr.: Harper & Row Publishers,I nc.

Frisinger, H. H., 1977: *The History of Meteorology: to 1800*. AMS, Science History Publications, N. Watson, Academic Publications, Inc.

Gillispie, C.C. (ed.), 1970: *Dictionary of Scientific Biography*. Charles

Scribner's Sons.
Griffiths, J.F.,"A chronology of items of meteorological interest". Bull. Amer. Meteor. Soc., 1977, 58, 1058-1067.
Hughes, P., 1976: *American Weather Stories*. US Dept. of Commerce, National Oceanic and Atmospheric Administration. (NOAA-S/T 76-1922). Available from Supt. of Documents. Washington, DC.
Khrgian, A. Kh., 1970: *Meteorology, a Historical Survey.* Translated from Russian. Israel Program for Scientific Translations.
Kutzbach, G.,1979: *The Thermal Theory of Cyclones. A history of meteorological thought in the nineteenth century.* Historical Monograph Series. American Meteorological Society.
Mason, S. F. 1962: *A History of the Sciences*. Collier Books.
Moulton, F. R. and J. J. Schifferes, 1960: *The Autobiography of Science.* Doubleday & Company, Inc.
Ochoa, G. and M. Corey, 1995: *The Time line Book of Science.* A Stonesong Press Book, Ballantine Books.
Pearce, L. 1878: *Album of Science, XIX Century.* Charles Scribner's Sons.
Shapley, H. S. Rapport, and H. Wright (eds.), 1965: *The New Treasury of Science*. Harper and Row Publishers.
Shaw, N. 1942: *Manual of Meteorology: Vol.I, Meteorology in history*. Cambridge University Press.
Weaver, J. H. (ed.), 1987: *The World of Physics,* Simon and Schuster.

Simple experiments:

Allaby, M., 1995: *How the Weather Works.* The Reader's Digest Association, Inc.
Brown, R. J., 1987: *200 Illustrated Science Experiments for Children.* TAB Books.
Boren, C. F. 1987: *Clouds in a Glass of Beer. Simple experiments in Atmospheric Physics*. J. Wiley & Sons.
Boren, C. F. 1991: *What light through younder window breaks.* J. Wiley & Sons.
Kikoin, I. K., 1985: *Physics in Your Kitchen Lab.* Mir Publishers, Moscow.
700 Science Experiments for Everyone. Compiled by UNESCO, 1958. Doubleday & Company.

Chapter 1, BASICS:

Gedzelman, S. T.: "The sky in art", Weatherwise, December 1991/ January 1992.
Neuman, J.: "Meteorological Aspects of the Battle of Waterloo". Bull. Amer. Meteor. Soc., 74, 3, March 1993, 413-420.
Williams, R.: "The divine wind". Weatherwise, October/November 1991, 11-14.

Chapter 2, AIR:

Allen, O. E., 1983:*Planet Earth: Atmos phere.* Time-Life Books.
Duffy, P. B., Comments on "Global

warming: a reduced threat?". Bull. Amer. Meteor. Soc., 74, 5, May 1993.

Ihde, A.J., 1`964: *The Development of Modern Chemistry*. Harper and Row.

Jackson, D. D., 1981: *The Aeronauts*. Time-Life Books.

Photzer, G.: "History of the use of balloons in Scientific Experiments". Space Science Reviews, vol. 13, no. 2, June 1972.

Chapter 3, PRESSURE:

Middleton, W. E. K., 1964: *The History of the Barometer*. Johns Hopkins.

Chapter 4, HEAT:

Middleton, W. E. K., 1969: *Invention of the Meteorological Instruments.* The Johns Hopkins Press.

Chapter 5, LIGHT:

Fauvel, J., R. Flood, M. Shortland and R. Wilson. 1994: *Let Newton Be:! New perspective on his life and work.* Oxford University Press.

Greenler, R. G., 1980: *Rainbow, halos, and glories*. Cambridge Univer sity Press.

Gleenler, R. G., "Laboratory simulation of inferior and superior mirages". Journal of the Optical Society of America, vol. 4, page 589, March 1987.

Minnaert, M.: *The Nature of Light and Colour in the Open Air*. Dover, 1954.

Walker, J., 1977: *The Flying Circus of Physics with answers*. John Willey & Sons.

Chapter 6, MOISTURE:

Whipple, A. B., 1982: *Storm*. Time-Life Books.

Middleton, W. E. K., 1965: *The History of the Theories of Rain and Other Forms of Precipitations*. Franklin Watts, Inc.

Chapter 7, FORCES:

Rogers, E. M., 1960: *Physics for the Inquiring mind*. Princeton University Press.

Chapter 8, MOTION:

Gleick, J., 1987: *Chaos, Making a new Science*. Viking.

Friedman, R. M. "Nobel Physics Prize in perspective". Nature, vol. 229, 27 August 1981, 793-798.

Lorenz, E. 1993: *The Essence of Chaos.* University of Washington Press.

Platzman, G. W.: "A retrospective view of Richardson's book on weather prediction". Bull. Am. Meteor. Soc., vol. 48, no. 8, August 1967.

Thompson, P. D and R. O'Brien, 1969: *Weather.* Time-Life Books, New York.

Wiin-Nielsen, A.: "The birth of numerical weather prediction". Tellus, 1991, 43AB, 36-42.

CREDITS

Design, layout and all line illustrations in this book were prepared by the author using a "Macintosh" personal computer. Images of scholars in biographical notes as well as other historical illustrations were obtained from the History of Science Collection, University of Oklahoma Libraries in Norman, Oklahoma.

The figures and quotations listed below were obtained as follows:

page 10 - top: drawn after the 19th century Japanese painting by Katsushika Hokusai, National Museum in Tokyo,

page 16: drawn after the 17th century frontispiece engraving by Sebastien Le Clerc in "Memoires pour Servir a l'Histoire Naturalle des Animaux" (1676) by Claude Perrault,

page 18: drawn after Figures 1 and 2 on page 74 in a treatise by Otto von Guericke, "Nova Experimenta di Vacuo Spatio" (1672),

page 32: drawn after illustrations in "Traite Elementaire de Chimie" (1789) by A. Lavoisier,

page 36: drawn after J. L. David's painting in Metropolitan Museum of Art, New York,

page 37: drawn after the illustration reproduced in "Planet Earth: Atmosphere", Time-Life Books, 1983, page 29)

page 44: redrawn based on illustration which was reproduced in a journal "Space Science Reviews", 13, 2, 1972, page 205 (redrawn here by permission of Kluwer Academic Publishers). The original illustration appeared in "The Illustrated London News", 39, 1861.1983. Guy-Lussac on the left is holding a flask, Biot on the right is holding a thermometer - barometer.

page 50, top: redrawn based on illustrations in "Dialogues Concerning the Two Chief of the World" (1638), by Galileo Galilei.

page 50, bottom: redrawn based on illustrations in "Cursus Philosphicus ... comprobata" (1653), by Emanuel Maignan.

page 53, bottom: redrawn based on illustrations in "Saggi di naturali esperienze fatte nell'Accademia del Cimento" (1667),

page 58: both sketches were redrawn after engravings (plates XI and XII on pages 104 and 106) in a treatise by Otto von Guericke, "Nova Experimenta di Vacuo Spatio" (1672),

page 66: reproduced after engraving on page 54 in a book by J. Glaisher, C. Flammarion, W. De Fonvielle and G.

Tissandier, "Travels in the Air", London, 1871,

page 71: drawn after a fresco by Gaspare Martellini, Museo Zoologico "La Specola", Florence,

page 80: Papin's digestor drawn based on a plate on page 69 in "Les Merveilles de la Science" by Louis Figuier (1867-1870),

page 92: redrawn based on Plate LXII in "The Complete Dictionary of Arts and Science" by T. H. Croker, Th. Williams, and S. Clark (1765),

page 111: drawn based on a woodcut from Descartes' "La Dioptrique", 1637,

page 118: drawn after a woodcut (plate XXXI) in "The Complete Dictionary of Arts and Science" by T. H. Croker, Th. Williams, and S. Clark (1765),

page 148: after a woodcut on page 168 in Descartes' "Les Meteores", (1637),

pages 170 and 182: the photographs on these pages were provided by Mr. John Elsworth,

page 172: drawn based on Bergeron's sketch published in 1972 in Journal de Recherches Atmospheric, 6, 49-53,

page 224: drawn after Plate 73 in Reynolds' 1883 paper, in Phil. Trans. R. Soc. Lond., pp. 935-982,

page 232, bottom: after Fig. 173 in Thomas Sprat's "History of the Royal Society", 1667,

page 232-top: based on illustration in the first volume (page 14) of "The Antiquities of Athens" by James Stuart and Nicholas Revert (1816),

page 257: the photo of a tornado was made by Dr. J. Straka. The remaining photographs in the book were taken by the author,

page 272: The sketch of the supercomputer was drafted by Sue Weygandt,

page 275: the radar image of a tornado was provided by Drs. J. Wurman, J. Straka, and E. Rasmussen,

page 276: the satellite image obtained from http://web.ngdc. noaa. gov/dmsp/image/hurricane/ erin0801.gif.